Featherstone's Complete Wargaming

Featherstone's Complete Wargaming

Donald Featherstone

DAVID & CHARLES
Newton Abbot   London

(p.1) Beautifully painted 54mm figures of Wellington and Staff

(p.3) Part of the impressive reconstruction of the battle on display at the Panorama, in the village of Waterloo, Belgium

British Library Cataloguing in Publication Data

Featherstone, Donald, 1918-
 Featherstone's complete wargaming.
 1. Wargaming
 I. Title
 793′.9

 ISBN 0-7153-9262-X

Typeset by Typesetters (Birmingham) Ltd,
Smethwick, West Midlands
and printed in Great Britain
by Butler & Tanner, Frome
for David & Charles Publishers plc
Brunel House   Newton Abbot   Devon

CONTENTS

INTRODUCTION

All military history, the entire spectrum of war, can be wargamed and in so doing confers undoubted pride and pleasure in commanding an immortal army of any chosen period. To do this requires certain artefacts, the first and most essential being model soldiers and this book is exclusively devoted to tabletop battles with these military miniatures. Next, a battlefield set up on table or floor, embellished with all the usual topographical features like hills, rivers, roads, villages and woods. Poised militantly upon it the inert mini-warriors await the breath of life that can be given in only one possible way, but hands however willing can only manoeuvre them when prompted by accepted guidelines and rules. An opposing commander is required – it is possible to play 'solo' – against yourself – but far better for someone of similar enthusiasms to be on the far side of the battlefield, doing their best to grind your army into the dust. But that is just the beginning, the bare bones that have to be systematically fleshed by the contents of these pages.

Begin by reading Chapter 1, 'Wargaming for Real', which, if you wish, can open the door to the absorbing interest of a lifetime, bestowing a fascinating march through centuries of warfare – from the chariots of Ancient Egypt at Kadesh in 1215BC to the Fallschirmjäger's gliders at Fort Eben Emael on 10 May 1940. It is unlikely in the beginning that the wargamer will prossess all the specific armies for the numerous historic battles described herein – however many lend themselves to being taken 'out of period' and can be fought in different uniforms. En route, this military miscellany can be dipped into for other aspects – the effects of weather on warfare, the 'fog of war', or the simulation of surprise which, grafted onto the game, will add colour and excitement. And let there be no mistake, everything written in this book is about a *game* – it is not incitement or training for real war, nor a solace for frustrated ex-Sergeants or Generals! Rather it emphasises the worst part of war as miniature metal casualties replace real ones.

The hobby of wargaming reveals innumerable facets and the selection offered here may be noteworthy for what has not been included. Nevertheless, *anything* said or written about wargaming can only be further stimulus to an already flourishing pastime that is all things to all men.

Above all, it belies its militant name by encouraging lasting friendships.

Assyrians

25mm Napoleonic Battle (Richard Ellis Collection)

1
WARGAMING FOR REAL

This is a 'key' chapter, linking together the 'realism' aspects that form a progressive theme throughout this book, and is by way of being the instruction-booklet explaining how it all works, aimed at:–

1. The practising wargamer wishing to improve his current campaigning and/or embark upon an alternative or second period.

2. The beginner, perhaps inspired by the hobby's colourful features portrayed in these pages, who wants to learn to wargame.

He (or she) can study the catalogues and buy an army; laboriously paint it; build a tabletop battlefield, and begin to fight battles, either 'solo', or against an opponent. However, whatever period of warfare is chosen, the resulting miniature battle will bear precious little resemblance to reality, because its tactics and manoeuvres will be inspired by contemporary knowledge and hindsight rather than to the knowledge of mobility, ranges, style-of-fighting, weapon-potentialities that would have been available to historical commanding-generals or even the man-in-the-ranks.

To remedy this, the principal fighting forces of military history are described, detailing uniforms, weapons, equipment, fighting-style and tactics, plus historical background. This provides the essential knowledge without which neither authentic nor realistic wargaming is possible. To round-off, each section concludes with a detailed description of an historical battle featuring the warriors under review, plus advice and instructions on refighting it as a tabletop wargame. From it can be gleaned invaluable information on the respective fighting-styles and tactics of the antagonists, their leaders and leadership, morale, etc, etc.

Thus armed, the wargamer can now be confident that the opposing forces manoeuvre and fight in an historically accurate manner. With so much information at his disposal, the wargamer might feel as though he is turning to the last page of a thriller to discover 'who done it', because both winner and loser are revealed. Slavishly following on the wargames table the precise course of events will probably produce the same result – which is all against the essentially competitive element of wargaming, being a demonstration or pageant-in-miniature. This is avoided by a variety of built-in 'ploys' – Military Possibilities; Chance Cards; Commander's Classification, etc, – which bring the exercise back to being a game, its result hanging in the balance.

The idea of refighting the famous battles of history, reproducing Waterloo, Gettysburg, Alamein, etc, on the wargames table, is a most attractive proposition. Unfortunately it is almost impossible to put it into practice with any degree of realism.

However, military history abounds with conflicts that are highly suitable for reproduction on the wargames table – battles that involve small numbers – so that realistic scaling down is possible. This is important because, if the battle is to bear anything more than a titular resemblance to the original, the tabletop armies must represent an accurate proportion of the original force both in numbers and types. It is possible for one man on the wargames table to represent as few as five men in real life, and some battles can be fought on a man-for-man basis. The size of the topographical features of the historical battlefield must lend themselves to reproduction on the wargames table, so choose those that are suitable in area and frontage for this purpose.

It is doubtful whether wargames will ever give one profound military insight, but the wargamer may gain an understanding of the problems of the commanders in the field and a glimpse of the military thinking of the time by refighting each battle in the correct tactical manner, using the formations and weapons of the day. The purpose of this book is to discuss these factors and

Before a big crowd, 165th Anniversary Battle of Waterloo wargame fought in a London hotel in June 1975 (Figures and terrain by Mike Willmore)

Plan a Wargame No. 2
The worst feature of wargames is the chopping and changing and the butterfly-mindedness of the commanders. Every tabletop threat seems to call forth an immediate reaction, often in the very same move. This could not happen in war and occurs in wargames partly because one player acts as all the generals, all the colonels and all the captains in the one army. Each wargamer sees the terrain before the battle but rarely writes plans or orders even in note form, preferring to dream up a few rather short-term objectives in the first place, after which he keeps a weather eye open and plays the game 'off the cuff'.

With a group of five players, it is possible to fight a wargame in a far more realistic fashion. The set up consists of an umpire and four players, two on each side, one of whom would be the general responsible for overall planning. The wargame table would be erected in one room, whilst each general would have his headquarters in another room. Here he would make his plans and give out his orders, working on the usual data as to ground forces available. These orders would be passed to the umpire who would then give them out to the two Junior Commanders, whose role would be to move the armies in accordance with the orders of his Senior, reporting any information that he felt to be important but he must not talk to his superior. The senior general, normally stationed at his headquarters, would be allowed to come to the rear of his side of the wargames table and review the situation at the end of each move. Should he wish to give out fresh orders then these have to be transmitted by a note via the umpire.
continued

to suggest practical methods of simulation that will produce an accurate, realistic and enjoyable facsimile of the original battle.

Except when being purposefully staged as a demonstration, wargaming is a competitive affair and selected battles should present the challenge of reversing their result. The tabletop battlefield is a miniature replica of the historical arena, and the armies conform to the original forces so that the table-top reconstruction of an historical battle resembles as closely as possible the events that occurred on the battlefield itself. It is pointless to construct a terrain resembling an historical battlefield and then let loose upon it a host of infantry, cavalry and artillery whose general milling about bears no relation to the original conflict. Every aspect of the battle has to be considered in its correct context, and in chronological order, so that it can be simulated and affected by the fluctuations and fortunes of war, without radically departing from the Military Possibilities of the day. The wargamer should consider what might have occurred in conjunction with what did occur and, while following the original course of events, be allowed some leeway, but they must be reasonable alternatives, or the reconstruction will lose its authenticity and become an ordinary wargame.

If the battles are so reconstructed that both armies are of historically accurate scaled-down strength, pursue the same tactical plans, use the same weapons and fight in a manner of their day, it is highly likely that the table-top encounter will follow its historical course. If not, the rules controlling the wargame may lack balance, or one wargamer, perhaps better versed in military tactics than his opponent, is manoeuvring his armies with a military hindsight denied to the real life commander of long ago. Wargames armies must not be tactically manoeuvred in a manner far beyond the knowledge and comprehension of their time. Such restraint is difficult for the twentieth-century wargamer, with his knowledge of later military actions, and his awareness of the historical course of the battle.

Each phase of a battle should be considered carefully, while seeking possible alternatives to the historical trend of events. These alternatives will be called Military Possibilities, which may be defined by reference to the Battle of Waterloo, for example. If Blucher had not arrived to aid Wellington, what might have happened? You can work it out on the wargames table!

Military Possibilities
Under analysis, every battle reveals stages when certain moves or tactics point the path to eventual victory or defeat. At each stage there are possible alternative courses of action to that taken, and some of these logical Military Possibilities may well lead to a complete reversal of the result. In some cases the course of action indicated by a Military Possibility results in a more reasonable and credible result than occurred on the historic field, but they must never be considered as excuses for indulging in whims and fancies, merely to 'see what happens'.

Like real war, wargames are competitive contests influenced by luck, which is represented on tabletop terrains by the throw of a dice or the turn of a Chance Card – simple methods of simulating the ebb-and-flow and the fluctuations of war.

Military Possibilities can be utilised as a means of encouraging the wargamer representing the defeated commander cheerfully to accept his substandard role by giving him an outside chance of reversing the result, without radically altering the historical course of the battle. They should be restricted, therefore, to relatively minor aspects that may bring about increasingly interesting tactical twists. Abounding in all battles, Military Possibilities bring colour to the wargames table in proportion to the ingenuity of the wargamer.

The Junior Commander would be allowed minor tactical discretion. It would be for the commanding general to lay down objectives rather than to dictate where, for example, units should be in column or in line.

Chance Cards

Closely akin to Military Possibilities, Chance Cards prescribe unexpected events that, by introducing pleasant and/or unpleasant factors, can materially affect the course or even the outcome of a battle. Their worded instructions pose mental, physical and tactical eventualities that require compliance by the commander drawing the card. They are useful, for example, when a wargamer needs to know whether a courier has arrived, and at what time, for they will state whether or not the man has been delayed (perhaps by guerilla attack), or if his thoroughbred

Re-fought Battle of Barnet, Wars of the Roses 1471

11

horse has brought him across country at a faster rate than expected.

Specific sets of cards can be designed to cover the eventualities peculiar to a battle. They may, for instance, control the death of the Imperialist cavalry commander Pappenheim at Lützen, which affected his cavalry so much that their promising counter-attack faded away, taking with it the Imperialists' last chance of saving the battle. Here is a set of Chance Cards to cover the numerous possibilities in this situation, bearing in mind that what occurred in fact must seem the most likely eventuality.

Card No 1 – Only the front rank sees Pappenheim fall, but it is forced onwards by the press of horsemen behind.

2 – Both front and second ranks see their leader fall and involuntarily ease up, so dissipating the impetus of their charge.

3 – Seeing Pappenheim fall, his cavalry force reins up and halts for a game-move.

4 – Seeing Pappenheim fall, his cavalry force reins up and halts for two game-moves.

5 – Seeing Pappenheim fall, his cavalry force reins up but, within half a move, is urged on by the entreaties of the second-in-command, now its leader.

6 – Realising that its leader has been killed, Pappenheim's cavalry force slows to a trot, thus losing the bonus effects of their charge.

7 – Realising that their beloved Pappenheim has been killed, the cavalry force carries on but with a reduced morale status.

Every battle of military history presents situations with opportunities for adding interest, and often realism, by the use of Chance Cards. The imaginative wargamer, indeed, may need to show restraint in their use at times!

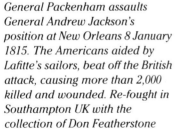

General Packenham assaults General Andrew Jackson's position at New Orleans 8 January 1815. The Americans aided by Lafitte's sailors, beat off the British attack, causing more than 2,000 killed and wounded. Re-fought in Southampton UK with the collection of Don Featherstone

Classification of Commanders

If both forces are equal in strength, morale, equipment, position, mobility, etc, on the wargames table or in real warfare, victory will go to the side with the best commander. On those occasions when a force is defeated by a numerically weaker army it is generally because the commander of the weaker side possesses outstanding tactical ability and can inspire his men to exceptional heights. Failing that, he may just be the better commander on the day. When wargaming, the ability of a commander is more or less that of the wargamer representing him, but this factor is balanced by giving those leaders who were historically superior an 'above average' classification, whereas known incompetents are classified 'below average'.

A massed French attack goes past the farm of La Haye Sainte in a re-fought Battle of Waterloo (figures and terrain by Peter Gilder)

On the wargames table the troops of an 'above average' commander display higher morale and better fighting qualities than those unfortunates led by an 'average' or 'below average' commander. This differentiation is achieved by adding or subtracting from the dice scores that control morale and fighting ability. For instance, troops led by an 'above average' commander add, say, 1 to their dice, troops under a 'below average' commander deduct 1, and those men led by an 'average' commander do not alter their dice scores.

An 'above average' commander, by being allowed to be 'himself', is granted greater flexibility of movement and, within certain limits, can employ tactics of another period and issue fresh orders at the start of each game-move; the 'below average' commander is compelled to manoeuvre and control his army strictly in conformation with its known style of fighting, and his troops must stick rigidly to their initial orders until they are disorganised by some forced reaction, such as a disastrous morale rating. The 'average' commander is allowed, within certain limits, to alter characteristic fighting patterns while controlling and manoeuvring his force in accordance with its known style of fighting, and is permitted to change his orders every third game-move.

No wargamer likes to follow rigidly a course of action that history tells him will bring defeat, particularly when he has a marked numerical superiority, so, to save his face in such a situation, the original commander may be classified as 'average' or 'below average'.

If a system is used where, at the start of the battle, orders have to be written for each army or group, the grading of commanders can be reflected by ruling that those armies or groups under 'below average' commanders must conform rigidly to their initial orders until they are disorganised by a forced reaction (such as a low morale rating). 'Average' commanders can write orders to carry through three moves of the game; at the conclusion of the third move fresh orders may be written if circumstances have not already altered the original instructions. 'Above average' commanders may write orders at the beginning of each move of the game.

Rorke's Drift 1879 (re-fought with Willie Figures)

Plan a Wargame No. 1

1. Two opposing armies are selected in the usual manner and a wargames terrain is laid. Each Wargamer is given a Commander-in-Chief plus a Corps Commander for each of the corps in which he intends to divide his army. Armies may be divided into two, three or four corps as desired.

2. In most real life battles, one army attacks while the other defends. This is not always the case on the wargames table, when both forces go gaily and militantly forward to meet each other. In this particular method, it must be decided which army is to attack and which army is to be on the defensive. The defensive army is then laid out in its desired dispositions not further forward than the half-way mark on the table. The attacker's forces are laid out in their Corps formation on his base line.

3. The attacker now nominates one Corps (together with supporting troops) with which he intends initially to attack. He lists the units in this Corps in the usual manner and briefly writes on his movement chart what he intends them to do. The defender, who has already laid out, plays the role of an interested spectator at this stage.

4. The attacker now discloses to the defender the moves that he has written down for his first Corps. The defender has to decide at this point whether he intends counter-attacking with his Corps (and other units) within reasonable reach of the enemy's first Corps. He is not permitted to pick one unit out of the first Corps to attack but must consider the group (including supports) as a whole. If the *continued*

An alternative is to allow all commanders, whatever their rating, to write their orders at the beginning of the game, but the lower the commander's grade, the greater number of moves he must take before he alters his orders, while the 'above average' commander may change his instructions much more rapidly.

An interesting (and sometimes amusing) method of simulating the classification of commanders is to bestow the mantle of the 'below average' commander upon novice wargamers and the hat of the 'above average' commander on an experienced or veteran wargamer.

Morale

Morale concerns the discipline and confidence of troops, both collectively and individually, and was believed by Napoleon to be three times as important as physical factors. In most well known battles the morale of the troops played a major part in deciding the outcome of the battle. Thus, there must be adequate simulations of this intangible factor, which causes men suddenly to break, or to rally and beat a force much larger than themselves, in spite of being attacked in flank or rear and having lost their officers. Inevitably, the morale rules used in wargames of any period bear a basic resemblance to each other, because the factors that frighten or stimulate soldiers have not really changed since the beginnings of time, though they are much affected by such variables as confidence in a commander, high standards of training and discipline, and battle experience.

Military history reveals many occasions when smaller forces were victorious because their morale was higher than that of their opponents. On the wargames table careful manipulation of the comparative morale status of opposing troops makes possible battles between forces with marked numerical disparities.

Morale factors feature prominently in the rules that control table-top wargames. The little mindless lead or plastic figures need to have intelligence and emotion bestowed upon them, to accord with specific times and circumstances. This can be reflected by troops starting the battle with an 'average' state of morale which, according to the fluctuations of combat, will rise or fall. For example, if the average morale status of a unit is represented by the figure 6, 1 may be added for the support of a friendly unit on flanks or rear, and another 1 if the Commander-in-Chief is with them, making a total of 8. You may deduct 1 from the total if the unit loses a quarter of its men, another 1 if they come under artillery fire, 1 more if they have to move back, and 2 if, in the last game-move, they lose a melee – leaving a final total of 3. Then a dice is thrown to represent those fluctuations of fortune that are an inevitable adjunct to war. If the score is 3, this figure, added to the original 3, indicates that the unit had maintained its morale value of 6 and is still in first-class shape. If the dice score is 2, then the unit's morale will fall below par, causing them to withdraw; and a dice throw of 1 will reduce the unit's morale to such a state that they may well run away from the field.

Time Charts

Many battles include some 'surprise factor' that needs recording, so that a check can be kept, for example, on the anticipated time of arrival of a flanking force. A Time Chart programmes such vital factors beyond dispute. The Chart must include those manoeuvres whose timing is an important feature of the battle; each of them, or stages in their accomplishment, should be treated as a move in the game.

A Time Chart is vital in keeping check on off-table moves on a map, where different forces are moving along various routes or attempting outflanking movements that will bring troops on to the tabletop battlefield at some intermediate stage in the conflict. It is almost impossible to retain control of such factors without a Time Chart.

Keeping in touch with detached portions of a force might require their commander, perhaps unaware of their exact location, sending messengers, whose progress must be recorded on the Time Chart. Thus, the messengers' exact time of arrival is known, and the unit to whom they are bringing orders cannot react until these orders are received. The non-arrival or delay of orders provides Military Possibilities that can realistically alter the course of a historical conflict. Any situation where forces are attempting a manoeuvre that will bring them on to the tabletop battlefield at some intermediate stage in the conflict requires recording on a Time Chart. If a commander sends messengers imploring aid, then the progress of these messengers, their exact time of arrival, and the subsequent time reaction of the unit receiving the message, must be recorded on a Time Chart, as must the progress of a relieving force.

defender decides not to counter-attack at this stage, then the attacker goes through the same procedure with his second corps, again disclosing its movements after they have been written down and again the defender has the chance to counter-attack. This procedure is followed for the third Corps, if there is one. If, at the end of this first round of moving, the defender has not intimated that he wishes to counter-attack, then the attacker carries on in the same way for the second round. This occurs until the defender exercises his right of counter-attacking.

5. When an attacking force is counter-attacked, both Attacker and Defender consider all the prevailing situations according to the proximity (in the light of their respective move-distances) of all other units. Allied units of both Attacker and Defender may move or fire as soon as action commences IF they can make contact in a single move or are within range of musketry or artillery.

6. When a round of firing has taken place and all melees are concluded then the attacker may move his other Corps to conform to the situation as it now exists. The battle is carried on in this manner with both Wargamers endeavouring to maintain a realistic continuous ebb and flow of combat. When accompanied by fairly strict morale rules which bring about a constant breaking and rallying of units, this method of wargaming, although difficult to maintain and requiring considerable give and take from both Wargamers, is realistic and demanding.

NOW YOU SEE THEM, NOW YOU DON'T!
Surprise and Concealment on the Wargames Table

From a wargaming point of view, SURPRISE is ageless and soldiers will be no less shocked today when a tank suddenly emerges in a village street than were Ancient Egyptians at the sight of Hittite chariots coming from behind a clump of trees. Thus, reactions simulated by rules must be similar – and equally devastating! Similarly, when studying accounts of historical battles through the Ages, the mode of surprise conceived by a Greek or Roman commander can be modified and utilised in any desired later period of warfare – as it often has been in real life.

Perhaps the greatest surprise in military history was the Allied invasion of Normandy of D-Day 6th June 1944 – too vast for tangible reconstruction on a wargames table. However, described elsewhere in these pages as a wargame is another, albeit minor, surprise operation of World War II – the German glider assault on the Belgian Ford of Eben Emael on 10th May 1940. History abound with examples of military surprise – a classic Medieval operation, described in these pages as a 'A Battle For All Seasons,' was Derby's attack on the French at Auberoche in 1345.

A certain degree of concealment and uncertainty can be introduced into a wargame by a little pre-planning. For example, let each general pick an army by means of a point system of 1 point for an infantryman, 1½ points for a cavalryman and 10 points for a gun – each general to have an army of his own choice to a prearranged total. Next throw eight dice and add to your forces additional troops to the points value of the total dice score. Decide which force is defending and which force is attacking and the defender will then lay out the tabletop terrain and arrange a curtain across the middle of the table. Decide from which side of the table the attacking force is to come and then give the invading general a rough sketch map of the defender's half of the field. The invading general will also have the advantage of moving first in an alternate-move game. Arrange the forces on the table on either side of the curtain, both sides being allowed to lay down within 12″ of the curtain. This makes it possible for units to be within artilery range but not within cavalry-charging distance or infantry-musketry range. Further modifications can be embodied in the laying out of your forces – units behind woods or buildings or on reverse slopes need not be laid down but a note must be made of their position and strength. Units in houses and woods can be represented by one man. All other units, including those on the edge of woods and in a position to fire, have to be set out completely, and all concealed or partly concealed units have to be set out as soon as they move out from cover or open fire.

Without surprise there is a lack of spice in a wargame and tactical surprise can only be brought about by concealment in some form or other, 20mm or 30mm metal soldiers cannot be hidden behind model trees or hills made of plasticine because the opposing general is probably a man six feet tall who can see everything! Cards or scraps of paper are unrealistic. Ruling treble-distance charge-moves (over a period of three game-moves), and allowing the charge to change direction any time before the final approach, enables your dispositions to give no clue to your future plans. After three moves each player is then given one move to adjust his forces before actual fighting can begin. If, by some chance, forces do not contact each other at the end of the agreed number of marked moves, then they are placed on the table as they were marked at the conclusion of that move.

There are many variations to this programming technique; for example, one can use columns coming in at map-move 1 with other columns not being programmed until map-move 5 or so. The system enables a player to make wide enveloping movements which are extremely difficult to make in the conventional game because his opponent can observe it as it starts and easily seek a counter.

While victory frequently goes to the big battalions, it can also be vastly influenced by shrewd tactical moves or the mere fluctuations of fortune. Surprise movements of bodies of troops so that they suddenly arrive upon the rear or flank of the unsuspecting enemy are difficult to simulate on the wargames table, but such movements charted on a map enable forces to be manoeuvred so that the enemy is unaware of their intention; and even their existence, until contact is made.

A relatively simple method of simulating surprise and concealment is for both commanders to have their maps covered by a sheet of plastic upon which each move is marked with a chinagraph pencil. The umpire takes both plastic sheets and places them on top of each other, with his master map at the bottom, so that he can determine whether any troops are within visual distance of each other or have come into contact. This procedure is repeated at the conclusion of each game move, with the umpire checking both maps and handing them back without comment until a contact is made or forces become visible to each other. Then the troops can be placed on the battlefield as though they had come upon each other from behind concealing topographical features.

The countryside surrounding the area of the battle (in the centre oblong) can be drawn inaccurately on the map of the commander who is due to be surprised, so that he is unable to estimate the possibilities of an outflanking movement or the time it will take to reach him. On the other hand, the commander of the outflanking side can be given an inaccurate map, a Military Possibility that will affect his surprise move and give the original loser a slight chance or reversing the decision. If such steps are taken, it is advisable to have an umpire with an accurate map, for that will save a lot of arguments.

A reasonably successful method of surprise and concealment, again using maps that can be accurate or inaccurate as desired, requires the commander making the surprise move to work out, on a scaled map, the number of game-moves that it will take. He writes down –

(a) the unit or units involved (or total strength of force).
(b) route from A to B (starting point of move to point of contact).
(c) number of moves required.

If the force making the surprise move was visible in the battle, it will remain on the wargames table and will not move until the game-move in which it strikes home. If it was not visible to the enemy, it will move exclusively on the map, to be revealed and placed on the table when its presence was discovered. Using written movement orders and with a scaled map of the terrain, the commander perpetrating the surprise plots it for the required number of game-moves, writing instructions in progressive columns. On the completion of the requisite number of moves, the surprise force is disclosed by the war-gamer controlling it. He must declare if he wishes to terminate his concealed move before its completion, when he must move his troops up to the point they have reached. Surprise is then considered to have been lost, even if his troops are now technically concealed. A suspicious enemy commander may challenge, but his suspicions will need to be reasonably precise, as to direction and intention – the advisability of using an umpire is stressed. In the event of something happening on the wargames table that interferes with the surprise move, such as an enemy force crossing or positioning itself on the line of march, the move will have to be revealed

Surprise in South Africa – Boers ambush a British infantry patrol amid rocky outcrops on the veldt. (Figures, British are Stadden 30mm; Swedish African Engineers and Minifigs (kneeling)

and a decision taken as to which side is more surprised. This eventuality could take the form of a Military Possibility (see the chapter on Wargaming for Real). Perhaps the most successful and best known system of map-moving in a manner that can be successfully carried out by two players lacking the services of an umpire is the Matchbox method. The first essential is to grub around until a large number of empty matchboxes has been collected. They are then glued together into a sort of chest-of-drawers so that 36 matchboxes would form a chest 6 boxes by 6 boxes and numbered from 1 to 36 from the top left working across to the bottom right. Each box in this chest now represents a grid for the campaign map which is correspondingly gridded into 36 squares and similarly numbered. The lower numbers are at the north end of the map whilst the higher numbers are at the south. Each general has a set of plastic counters with numbers or letters marked upon them, each counter representing a part of his force. For example, counter A might represent a Brigade of Guard Infantry. counter D a Squadron of Cavalry and counter F a battery of guns. He notes on a separate chart against the numbered counters exactly what force each counter represents and then, having plotted his forces on his campaign map, puts the corresponding counter into the numbered box in the chest that represents the numbered square on the map in which that

Perhaps the greatest surprise possible in warfare is the ambush – here demonstrated by Apache Indians against a patrol of US Cavalry (Courtesy of Duncan Macfarlane, Wargames Illustrated*)*

formation has been disposed. Another method is to use small slips of card on which is marked the composition of the force it represents. As the forces move to another grid square on the map so the counters (or cards) are moved to the corresponding matchbox.

A certain code of ethics governs this almost ritualistic procedure – one general turns his back whilst the other slips his counters into the matchbox and no general who is a gentleman would even think of opening a matchbox unless it is to put a counter inside it! On opening a matchbox, should a general find that it contains one of his opponent's counters then he announces that a contact has been made and the necessary steps are taken to transfer the action on to the table-top battlefield.

There are certain refinements to this method, many of which will suggest themselves to the reader. For example, if part of the campaign map includes a large open plain (e.g. grid squares number 11, 12, 13, 14 and 21, 22, 23 and 24) where visibility exceeds 1 grid square, then for each force in the plain an additional counter is put into another pre-arranged matchbox (e.g. 21) covering the central part of the plain. By this means all forces that would in actual fact be visible to each other are known (via matchbox 21) to both generals. On the other hand, where a visibility barrier such as a range of hills, divides a grid square, then the card is not put into the appropriat matchbox until the map moves have taken it to the opponent's side of the hills. Although a certain degree of tactical surprise and apparent concealment is made possible by each commander initially drawing up a plan of the tactics he intends to use, giving a broad outline of the role to be played by each unit of his army. Both commanders have 8 playing cards, 2 being aces. Before each move a card is drawn, and the commander may alter a unit's alloted role if he draws an ace, but, if not, he is committed to his original plan. This method can be used in conjunction with the ratings of commandera, simulating the decisiveness of De la Rey and the 'woolliness' of Methuen at Modder River by giving the British 1 ace the Boers 3 aces per 8 cards. (This battle is featured in this book.)

Perhaps the simplest way of achieving surprise is for the host not to tell the visiting player the name of the battle he is fighting. For example, without mentioning the name of the battle the host could give the visitor the story of the events leading up to Cope's army forming up and facing south at Prestonpans, a course the visitor follows on the wargames table only to find, as did Cope, that the Highlanders have suddenly arrived on his left flank.

With wargamers towering in Godlike fashion over their table-top battlefield, the factor of surprise is most difficult to simulate, yet the recurring presence of this factor in warfare means that it must be represented if battles are to be accurately reconstructed. Only then can wargaming justify itself.

2
INSTANT WARGAMING
or How to Become a Tabletop General Without Wasting Time

Consider the virgin wargamer, stimulated by this book to take advantage of a time-warp and be transformed into the renowned commander of a famed army noted for its glorious historical victories – itching to smell powder with a comforting freedom from fear and danger, he doesn't know where or how to begin! Probably he is already aware of the need to know something of the armies, their uniforms and weapons, without realising how much more to it there is than that! Certainly he needs details of the break-down into Corps, Divisions, Brigades and Regiments – and their respective numerical strengths. He has to be familiar with the soldiers themselves, the regiments in which they serve, their uniforms, equipment and weapons – which includes knowledge of weapon-effectiveness, range, rate of fire etc, etc. He wants to fight his battles in the style of the period so that Napoleonic infantry do not unconsciously take on the modern attributes of World War II soldiers, so he has to be familiar with the tactics and style-of-fighting of his chosen era.

And the background to all this, the backcloth derived from historical causes and progression of the war, its principal battles and a lot more – that has to be considered too. He is a farseeing wargamer who chooses a period that is well documented, about which innumerable volumes have been written over the years, and are readily available through the Public Library. The more he learns the more likely are the rules he formulates to accurately simulate the desired historical warfare. He may not want to make a career in wargaming, but at least he is now aware of what he is taking on, and his required 'education' before qualifying as a capable Tabletop General. It will help if he has a model of what is required – so here is an 'Instant' manual of perhaps the best documented conflict in the history of warfare.

Were a wargamer's tabletop activities to be restricted to selecting a single war from the many that have plagued mankind since the beginnings of time, high on the list must be the American 'War Between the States', fought in 1861–1865 and well-known to most, if only through the film *Gone With the Wind*. It provides a fascinating start to a wargaming career, a large-scale and prolonged conflict that because of its technological aspects coupled with the extensive use of man-made defences was probably the major military link between the wars of yesterday and those of today. Despite being fought on a continental scale by large armies using relatively modern weapons, its

From an original photograph, this is an Infantry Private of the US Army in 1861. His dark blue fatigue jacket was standard campaign dress, and his weapon was the US rifled musket, Model 1861

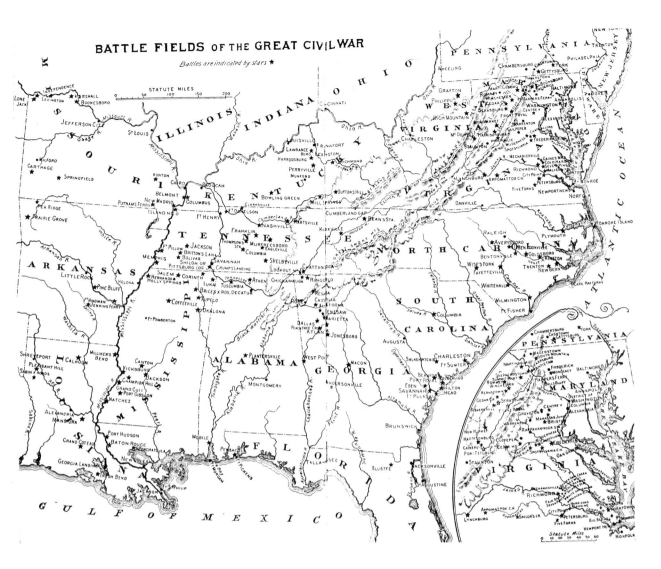

tactics and manoeuvres oddly reflected the colourful Napoleonic campaigns of half a century earlier – minus the gorgeous uniforms, of course!

What follows can reasonably claim to be just about all the novice needs to know to become a practising wargamer – except the model soldiers themselves. And even those essential items bear the 'instant' tag, because all the best hobby and model shops sell boxes containing about 40 HO scale 1:72nd plastic figures of Federal and Confederate infantry, cavalry and gunners with their guns and horse-teams – for less than £2 per box. Elsewhere in these pages are details of selecting from each box figures in similar positions and forming them into regiments, so that from about six boxes twelve 20-man or so regiments are formed – and you can't wargame much cheaper than that! Using these inexpensive figures includes another bonus in that they are moulded in plastic which is the basic colour of their uniform – in this case Federals in blue and Confederates in grey, so that they can be sent into action fairly rapidly by merely painting the faces and hands, weapons and the green base.

The American Civil War 1861–1865

Big in every aspect, this conflict was fought by citizen-soldiers in armies that progressively increased in size, firepower and standards of leadership. There were some 2,000 confrontations, of which 149 were considered sufficiently important to be called battles; mostly, they were fought in an area of 150 square miles (400 square kilometres), that was well populated, contained the two capital cities, and was notably marked by woods, valleys, rivers and railways. In both tactics and weapons it was the first conflict in history to be readily identified with present-day warfare, with railways and telegraphs playing important parts; steamships fought and transported armies; metallic cartridges were produced for the first breech-loading repeating rifles; the revolver, the machine-gun, and rifled artillery took their place on the battlefield, and wire entanglements were first used.

The twenty-two Northern States (including three Border States with divided loyalties) had a population of some 22 million; the eleven seceding States of the South had 5.5 million white and 3.5 million negro slaves; the available manpower ratio favoured the North by 5:2. When the War began, both sides utilised militia and poorly-trained volunteers on short enlistment terms, formed into regiments that went almost untrained into battle. Without veteran-soldiers, no regimental loyalties, and commanded by officers who knew no more about war than the man in the ranks, it is surprising they performed so well. Regiments were just about as good as their officers, and units without confidence in the men who led them lacked confidence in themselves – if the officers stood and displayed bravery then usually the regiment held, but with inadequate leaders the regiments wavered and broke.

Although the men who fought the Civil War displayed the highest qualities of bravery, fortitude and self-reliance, their lack of discipline frequently deprived them of the fruits of many hard fought battles. Without the officers to supervise

Surrendering and Prisoners-of-War

Wargamers are notoriously lax on the question of prisoners, preferring to fight to the last man simply because it is difficult to provide on the wargames table the human inducement to surrender. It has to be made worthwhile in the broad sense for a wargame 'general' to surrender a portion of his force, although in real-life the taking of prisoners is logical and highly realistic.

A possible inducement might be for each prisoner to require a 'keeper', in other words if twelve men surrender then it takes the same number of men out of the enemy's ranks to guard them. The tough commando or 'elite' soldier, probably thinner on the ground than his opponents, could take out TWO men as guards for each prisoner. The guards take the prisoners back to a specified HQ or to the baseline, 50 percent of them being able to return to the battle – but the surrender has taken away valuable men at perhaps a crucial moment in the battle. Commandos/Rangers might only need to provide one guard for every two prisoners; or it may be considered tactically expedient to refuse to accept (through dice-throws) surrender and shoot down the enemy.

The 1-for-1 (or 1-for-2) prisoner and guard method could be used in conjunction with an enforced surrender system, with double-strength attacking troops calling upon opponents to give in, who can fight it off by scoring 50 percent or more on the percentage-dice.

The Parrot Rifle – with a maximum effective range of 5,000 yards, this was one of the first rifled field-guns used by the US Army

Union Infantry advancing

training and lacking the national characteristics that produced the machine-like obedience of the European soldier, Johnny Reb and Billy Yank fought like demons but strayed like schoolboys.

At the beginning of the War, the armies of the South took the field equipped with nearly every military quality, with their cavalry particularly good because the Southern-white of every social class was brought up to ride and shoot. Perhaps the morale of the Southern soldier, defending his home against an invader, stimulated by early successes and led by an excellent and much-loved commander, was higher. In the face of great numerical superiority, or when tactically defeated, he knew he could hold his ground. The ragged, ill-shod and badly equipped Confederate army that marched towards Gettysburg in 1863, was a well-organised and trained formation that had not met a major reverse in two years of fighting – it must be recognised as one of the great armies of military history.

Throughout the long war the Northern soldier had to learn everything in a military sense; he did not know how to adapt himself to outdoor life and subsequently suffered badly from illness and disease. Perhaps the only advantage the Northern soldier had over his Confederate counterpart was an innate capacity for discipline. As the War progressed, the Federal soldiers of the Army of the Potomac, repeatedly out-fought, out-generalled and defeated, persistently bounced back under some new leader until at Gettysburg these city-dwellers and farm-lads from the East discovered a degree of morale grafted on them by gradually emerging corps and divisional commanders. In other theatres of war it was different – men from Minnesota and Iowa in the Army of the Tennessee possessed a proficiency with weapons that more than equalled the Southerners, besides being infinitely better at obeying their officers. In the West Union troops were mostly farmers with many of the ready-made attributes that go to form a capable soldier, so they too could match their Confederate opponents.

Both sides were similarly organised with the Regiment as the basic infantry and cavalry unit, known by their number and their State – the 2nd Vermont or the 7th Alabama. In 1863, the average Union regiment of infantry usually mustered about 425 effectives and was allowed to dwindle away through casualties and sickness until reduced to about 150–200 men, when it was broken up and, with the remnants of other units, formed into a new regiment. This practice destroyed Regimental loyalty and was better handled by the Confederates who kept existing regiments intact while feeding-in recruits; however their units were always well below strength. Regiments were grouped into brigades; three brigades formed a division and three divisions a corps; corps made up armies. Union army-corps and divisions were smaller than those of their opponents – 11 Confederate divisions equalling 17 Federal divisions. Tending to be a little stronger than Union regiments, a strong Confederate *infantry* brigade contained the same number of men as a weak *all-arms*

Federal Division, but it lacked the same fighting effectiveness.

In the Southern Armies, regiments of the same State were brigaded together and formations bore the name of their original leader, even after the command had passed to another. Federal corps were more likely to be known by numbers, while the Confederate formation bore the name of its leader. Armies were usually referred to by locality, such as the Union Armies of the Potomac, the Ohio and the Cumberland, and Confederate Armies of Northern Virginia, Tennessee and the West of the Confederacy.

There were distinctive flags for Armies, Corps, Divisions, Brigades and Regiments; they were carried into action by Colour-Guards who suffered heavy losses serving their purpose as rallying-points, maintaining morale, and indicating the unit's position and progress.

The Union were in a far better position to anticipate eventual victory through their almost unlimited supply of man-power, whereas the South could not greatly increase the numbers of their soldiers. But it was not numerical inferiority that defeated the barefooted and ragged Army of Virginia, it was lack of supplies – logistics were the great deficiency in Lee's generalship.

It has been suggested that not having won the War in its early stages, the Confederacy were bound to lose because they could never hope to improve upon the genius of Lee as a commander, whereas the Union, by a process of elimination of commanders, might throw up a leader who could approach Lee's brilliance – which occurred when U. S. Grant rose to the top. In fact, by capturing Vicksburg in mid-1863 – rather than the victory at Gettysburg at the same time – Grant won the Civil War for the Union because, as Sherman said: 'The possession of Vicksburg is the possesion of America.

Encounter between a Union horseman and a Confederate infantryman

The Leaders
In 1861, when the War broke-out, the best and most experienced officers of the United States Army offered their allegiance to the Confederacy, who held this superiority in leadership throughout the War. A unique aspect was that most of the opposing commanders were intimately acquainted with each others strengths and weaknesses, through their associations at West Point and in the Mexican War of some twenty years earlier.

During the course of War each side had at least three leaders who were more able than most commanders of the Napoleonic Wars. General Robert E. Lee of the South was perhaps the best military thinker and organiser of his time, although he did not begin the War as Confederate Commander-in-Chief, as at that time it was considered that the best leader in the South was General A. S. Johnston, killed at Shiloh in 1862. Alone of the generals on both sides, the eventual Federal commander General Ulysses S. Grant demonstrated the capacity to command small forces as well as large ones in battle under a great variety of conditions.

The Southern commanders at the beginning of the War –

Lee, Longstreet, Joe Johnston, Bragg and Forrest – were still commanding at the end; while the Union had thrown up a host of young aggressive generals like Sheridan Custer, Wilson, Upton and Kilpatrick, and Reynolds killed at Gettysburg and said to be '. . . the most complete soldier in the Army'. In addition to finding a General in Grant, President Lincoln also was blessed with Sherman, Thomas and Hancock, all first-class leaders who would have made their mark in any war. So too would General Thomas J. Jackson, killed at Chancellorsville in 1863, who won the nickname of 'Stonewall' at First Bull Run, where he displayed all the militant pleasure of an Old Testament warrior in the skill and cunning of warfare. His infantry, known as 'Jackson's Foot Cavalry', marched and manoeuvred rapidly so as to give him numerical superiority. His death and that of General J. E. B. Stuart at Yellow Tavern in 1864 – the loss of this renowned cavalry leader caused the Army of Northern Virginia to stumble along in a half-blinded condition – were felt by the Confederacy for the rest of the War.

The Infantry

The American Civil War saw tactics revolutionised by improvements in weapons, bringing about the disappearance of the solid infantry formations of past wars, and the old type of cavalry conflict where shock-action destroyed and scattered infantry. Firepower made frontal assaults impossible, forcing formations to disperse and become more flexible. More than in any previous war, it was a conflict where the soldier was ruled by firepower so that battles were frequently decided by one side being armed with superior numbers of modern weapons, rather than to any well-planned tactics of the generals.

Commanders, particularly those of the Union, were so obsessed with the outmoded tactics of the Napoleonic Wars that they lost battle after battle through persistently ordering forward columns of bayonets that were shot to pieces by concealed opponents. Sheer volume of fire prevented attackers coming to close quarters so that the bayonet lost its importance as infantry in mass-formation emerged from cover, to be torn to shreds by case-shot and shell, fired by concentrated and untouched enemy artillery. The enthusiasm and fervour of the charge, aroused by loud shouts and the wild 'Rebel Yell', by drums and bugles, frequently lost momentum and petered out in a fire-fight long before reaching the enemy line. From each opposing gun, an infantry attack over 1,500yd (1,370m) would have to take –

20 rounds of spherical case-shot as they covered the first 850yd (780m).

7 solid shot in the next 300yd (270m).

9 canister-shot in the following 250yd (230m), and

2 rounds of canister in the final 'double-and-charge' over the last 100yd (90m).

The infantry tactics of the latter years of the War saw waves of brigades or regiments attacking at intervals of 250–300yd (230–270m), allowing the second line comparative freedom from enemy musketry and able to act in close-support, with room for units to swing right or left if attacked in flank. Regimental attacks in columns of companies and divisional attacks in columns of regiments were delivered against a specific point in the enemy line or where restricted ground prevented deployment.

Hard experience forced skirmish-lines to grow heavier until taking up half the strength of the regiment, the remainder being held in line of battle as a reserve. Taking advantage of ground, skirmishers fought in open-order, often as much as 400–500yd (370–460m) in advance of their main formation as they sought to keep down enemy fire and disorder his ranks with their musketry. They were not so badly exposed as might be expected, because to use artillery against well hidden skirmishers laying down behind rocks and tree-trunks was akin to shooting mosquitoes with a rifle!

Moving into action, a regiment could fire with all its companies abreast in a long double line; sometimes one or more companies were held back as a reserve or formed on the flanks. Troops under fire but not actually committed were usually ordered to lay down; the officers remaining standing. Fire was often held until the enemy was so close that it was almost impossible to miss; loading by numbers and firing volleys steadied the men and made correct loading of their rifles far more likely.

Both sides quickly learned to throw up log-faced earthworks; the Confederate armies, struggling to match the fire-power of the North, showed great ingenuity in defensive warfare, particularly in the use of artillery and defensive fox-holes dug by riflemen. Lee was a master of infantry tactics based on troops under fire digging-in until their fox-holes and rifle-pits expanded into trench systems. The great Southern commander used this as a base for both defensive and offensive operations. In front of Petersburg both sides were entrenched sometimes only 50yd (45m) apart – a veteran of World War I would have felt quite

Colt Navy
Revolver
Cal .32

Remington
Army
Revolver
Cal .44

25

at home among the wire entanglements, listening-posts, bomb-proof shelters and rifle-pits of 1864.

Aware that an infantryman in a rifle-pit was the master of at least twice his own number of attackers, and that a regiment behind earthworks was equal to a division in the open, General Lee frequently divided his armies in the face of superior numbers, to create opportunities for decisive counter-attacks. Improvised breastworks were erected, the rear ranks carrying forward rails, rocks, logs and any possible form of portable cover to where the front ranks, with all the arms, were holding the line. These barricades escalated to become efficient fortifications with salient and re-entering angles taking every advantage of the ground and allowing deadly crossfire. Such defensive skill was matched by the soldier's quick and shrewd appreciation of his chances of success when ordered to attack strongly fortified positions. On numerous occasions veteran regiments refused to advance, leaving the assault to less experienced and more reckless units. Unlike the European soldier who was controlled by strict and blind discipline the foot-soldier of the American Civil War was quite willing to risk his life when there was a chance of success, but he refused to throw it away. The old soldiers, the veterans, advanced in short rushes, laying flat between volleys and taking cover whenever possible.

Without realisation or intention, the Confederate soldier became the first of the major armies of history to wear camouflage-coloured uniforms; worn for want of anything else, their butternut-brown homespun clothing reduced casualties by allowing concealment and surprise. The basic uniform of Federal troops was the dark-blue tunic and light-blue pants, with a blue forage-cap (kepi); Southern soldiers wore uniforms varying from the elegant to the ragged and often lacked shoes. When the War began, both sides fielded units garbed in brightly-coloured baggy pants and boleros, with fezzes and turbans, in imitation of the dashing French Zouaves of the recent Crimean War; some even wore kilts. Such fanciful and impractical clothing soon vanished after a few months of campaigning.

An infantry regiment could usually average some 2½ miles (4km) an hour including halts, and a brigade of four regiments each of 600 men covered about 1,000yd (900m) of road. In two-rank battle-order, they occupied about the same area. An infantryman took up a personal frontage of about 2ft (0.6m), so that the unit-front equalled the number of men divided by the number of ranks, multiplied by 2ft (0.6m).

Doing his duty in the ranks today but off home tomorrow to help in the ploughing, throughout the war both sides fought against the handicap of having thousands of men wandering off when they thought fit, some to rejoin the colours when they had fulfilled the home-task for which they had deserted, others never returned or rejoined other regiments to get a further cash-bounty. Nevertheless and despite their obvious lack of 'spit and polish', they were unparalleled in ingenuity and inventiveness.

Opposite:
At the beginning of the Civil War enthusiastic volunteers formed themselves into Regiments wearing fanciful uniforms owing much to recent military events overseas. Thus, the second picture shows an infantry unit wearing the popular 'Havelock' headgear named after the Indian Mutiny hero and worn by British infantry in that war. The third group are a Zouve regiment, wearing the typical bolero jacket of the French North African soldiers who had been prominent in the Crimea War of six years earlier. The first picture shows a group of 'Highlanders' possibly the New York 'Highlanders', formed of Scottish emigrants

Perhaps it is fair to claim they were a breed of soldier quite beyond anything the world had ever seen before, or probably will ever see again.

The Civil War was fought with a multiciplicity of small-arms – the North officially adopted some 120 different models of rifles, muskets and pistols, besides purchasing and issuing more than 4 million muzzle-loading rifles designed to fire the Minié-ball. The smoothbore muzzle-loading muskets of the war's early days was good at 50yd (45m) range; at 100yd (90m) a man stood a fair chance of being unhurt; at 200yd (180m) there was little risk of him being hit except by a stray ball. Basically, the standard infantry arm of the war was the Springfield Model 1861/2/4 rifle, a rugged simple weapon with an effective range of 500–600yd (460–550m) and deadly at battle-ranges of 200–300yd (180–270m). Its rate of fire was slow and three shots a minute was considered fast shooting.

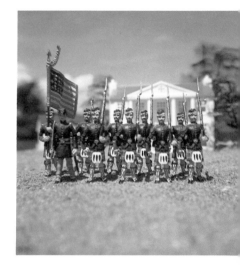

Training in marksmanship was almost unknown and many recruits went into battle without ever firing a single practice round, causing many a rifle to be fired unaimed into the blue. Nevertheless casualties increased as the war stretched its course because the smoothbore muskets of the 1861 infantry had become accurate and reliable repeating-rifles by 1864. Going a long way toward revolutionising warfare of the day, the Spencer rifle and carbine were the most widely used breechloading repeaters of the War – eight shots could be fired as rapidly as the lever could be worked and the hammer thumbed-back; in tests, it fired 21 shots a minute. It was this carbine that was the deciding factor in many of the great cavalry actions of the later days of the War. Then there was the Henry Repeating Rifle, with a higher rate of fire of a reputed 120 shells being loaded and fired in less than six minutes which caused Southern troops to talk of '. . . that damned Yankee rifle that can be loaded on Sunday and fired all week!'

Bayonets were carried by both sides but rarely used as a weapon – out of more than 7,000 wounded during Grant's Wilderness Campaign, only 6 were known as being wounded by sword or bayonet; more often than not it served as an entrenching-tool, can-opener or spit for roasting meat. Sheridan's battle charge at Stones River on 31 December 1862 was one of the few hand-to-hand struggles of the entire war; although bayonets were fixed before a charge, the attackers, brought to a halt by sheer fire-power, became involved in fire-fights before the bayonet could be used.

Officers carried the standard 'cap-and-ball' five or six chambered revolver of the period and traditionally most officers carried swords.

Uniforms

As well as the familiar blue and grey, there were many other colourful and fascinating uniforms. 'Zouaves', 'Guards', 'Chasseurs' all had their distinctive styles.

A Method of Simulating Infantry Attacking in Column

A column may be:-

a) A battalion column of 4 men wide by 10 men deep
b) A regimental column of 4 men wide by 20 men deep

Alternative columns may be formed of any number of men wide providing that the depth of the column contains more figures than its frontage.

An advancing column may fire during its approach, although the only people capable of doing this are those in the front rank and on the outside of the files. When a column takes casualties in the approach it must test morale to see whether it charges home. Should it fail to reach appropriate morale-rating, the column turns at the halfway point and retreats back to its starting point. It is probably more authentic to arrange Morale so that a line does not stand against a column unless they are very firm and of high quality. On the approach of the column, the defenders must test morale to see if they stand and face it (this can be done at the moment when the column reaches its halfway point). If their morale is not sufficiently high to stand, they retreat a specified distance in good order or in disorder, according to their morale rating. When the column reaches the enemy, it drives in 3in (7.5cm), pushing before it the same number of men (regardless of the number of ranks) as form the front of the column i.e. a column 4 men wide will push a gap 4 men wide in the enemy line which may also be 1, 2 or even 3 ranks deep. The first round of the melee takes place between the front rank of the column and the front rank of the men it has

continued

INFANTRY: Federal Regulars

The Regulation issue was dark blue kepi and four button coat, light blue trousers; all accoutrements were black leather, including the kepi peak, and on the kepi crown the corps insignia was displayed. The canteen was grey; the pack was black, the blanket roll red or light blue; bayonet scabbard was black with brass tip and handle fittings; all buckles, buttons and belt-plates were brass. NCO's had chevrons of light blue, and carried a short sword in place of the bayonet. The knapsack was often discarded and a rolled blanket carried over the left shoulder substituted. Many regular units, like the Iron Brigade of the West, adopted the black slouch hat in place of the kepi; use Confederate figures painted as Federals for this unit. Another famous regiment, 1st US (Berdans) Sharpshooters, wore green kepi, jacket and trousers, brown knapsack and light blue canteen.

There were colourful Zouave regiments, who can be adapted and converted from existing figures or, as there are so few of them, the expense of buying the specific metal figures might be justified.

INFANTRY: Confederate Regulars

Use plastic Federal figures with light blue kepi with dark blue band, grey jacket with light blue collar and cuffs, and light or sky blue trousers. However, the vast majority of Rebel infantry wore the slouch hat, in colours from fawn to black. The knapsack gave way to the blanket roll across the chest, and the uniform colour varied from brown or 'butternut' to many shades of grey. Texan units wore silver star (Lone Star) on hat. Trousers were often tucked into socks (made as for gaiters).

Louisiana Tiger Rifles wore a Zouave outfit. Baggy trousers were made from bed ticking (white with thin blue stripes). Short jacket was drab olive brown, with red trim. Shirt and cap red, cap with blue tassel. Gaiters white.

Infantry Flags

Union regiments had two flags; National colours, the stars and stripes, with the regiments name and number on centre stripe, and Regimental colours, blue with the name on a scroll beneath the eagle. Both will be 22mm (on pole) by 23.75mm. Confederate regiments had a Battle flag with 13 stars, 15mm square. Some also carried state or regimental colours (ie Texans, the Bonnie Blue Flag, white single star on blue ground).

The Cavalry

In the early days of the war the Confederate cavalry rode rings around the enemy, to raid, burn and loot almost without loss to themselves. In time, the Union produced increasingly proficient cavalry units until, always better equipped than the Southerners, the Federal mounted-arm overtook then bettered their Southern counterpart. The Northern cavalry had begun to assert and establish themselves by early 1863, with recruited plainsmen from the West and lively new leaders such as Pleasonton; Buford; Gregg; Kilpatrick; Farnsworth (killed at Gettysburg); Merritt; Custer; Wilson and, last but not least, Sheridan. During the last two years of the war none in the Army of the Potomac did more to overthrow Lee than the cavalry which operated in the Valley of the Shenandoah and at Petersburg.

At the start of the war, the twelve-troop organisation was adopted for Federal cavalry regiments, with a troop formed of 4 officers and 100 men; in 1863, cavalry-troops were reformed into formations of 3 officers and 82–100 men and battalions, usually of 4 troops, were formed. Union cavalry divisions had 2–3 brigades, each of 4–6 regiments. However, unit establishments were rarely filled and few cavalry formations on either side took the field up to their 'paper' strength. On paper, Southern cavalry regiments consisted of 10 companies or squadrons each of 60–80 troopers; Confederate cavalry brigades were made up of 4–6 regiments (depending on their strength) with cavalry divisions having up to 6 brigades, and 2 or 3 divisions forming a corps.

The ideal position from which to launch an attack was from the flank, although cavalry charges against weapons which, even in ill-trained hands could be devastating at 500yd (460m), were a crude form of suicide. Horsemen attacking over 1,500yd (1,370m) covered the first 620yd (570m) at the trot; 440yd (400m) at a manoeuvrable-gallop, and the last 440yd (400m) at the gallop-and-charge – which took them about 4½ minutes. During this brief period from each gun facing them they took 7 rounds of spherical-case; 2 rounds of solid-shot, and 2 rounds of canister!

Outstanding Confederate cavalry leaders were J. E. B. Stuart, Wade Hampton, Wheeler, S. D. Lee, Fitz-Lee, and the semi-guerilla cavalry commanders Bedford Forrest and Morgan. These last two led seemingly undisciplined and disorganised forces that, wearing no recognisable uniforms, must have resembled Boer commandos; varying in size from 500 men one day to 5,000 the next, the 'wanderers' became 'peaceful citizens' until they rejoined for the next raid. Spreading alarm and confusion in rear areas, these raiders tore up miles of railway-track and burned down Union supply depots. They were mounted riflemen who rarely fought from the saddle – Morgan's tactics included using a cavalry regiment drawn up in single rank with flank-companies (often on foot) to cover the whole front of the regiment; with one out of each group of four men holding horses in the rear or on the flanks, the regiment dismounted and deployed in files two yards apart in a line, attacking on foot at the double. A small body of mounted men were kept in reserve to act on the flanks, cover retreats, or press home a victory.

Civil War cavalrymen were armed with carbines well suited for cavalry use, being shorter and lighter than rifles, and with a high rate of fire as they were magazine-repeaters; accuracy was fair, 150–200yd (135–180m) was considered an effective range, although a fair percentage of hits could be made up to 500yd

pushed back, with those in the column being given some sort of bonus to represent the shock of impact. While is going on the remainder of the defenders have to curve back to conform, and in the second round of fighting those defenders who are in contact with the flanks of the column take part against those flanks, who also get a bonus for shock effect.

Unless the defenders push the column back after both rounds of fighting, the column is considered to have broken through the line and can move its full distance with remnants of line brushed aside on its flanks.

Rush's Pennsylvania Lancers

(460m). Other weapons were the sabre and the revolver.

For the first two years of the war, Confederate cavalry – an arm priding itself on being based on aristocratic traditions – performed dashing feats while glorying in the belief that the sabre was a more elegant weapon than the utilitarian revolver. At the same time, the Federal cavalryman, armed with the light cavalry sabre and a revolver, lacking the Southerner's natural propensity for horsemanship, developed 'dragoon' tactics, discarding the sabre and dismounting to fight on foot, using Spenser repeater carbines with deadly effect, reserving revolvers for close-quarter combat. It was the carbines of Buford's dismounted cavalry that saved the Union Army on the first day at Gettysburg. Eventually the Southern horsemen, acknowledging the need for more practical weapons, also abandoned the glamorous sabre and took to carrying large numbers of carbines, rifles, double-barrelled shotguns and revolvers – sometimes each man carried as many as four such weapons.

Thus, both Federal and Confederate cavalry, fighting with weapons that could be used either from the saddle or on foot, fulfilled a much freer role than the traditional European cavalryman, replacing the clash-and-parry of mounted charges by dismounted firefights that raged throughout the day. If necessary, they were expected to fight on foot, to seize and hold ground until the infantry arrived, but without being just mounted infantrymen. At times, in one and the same action, they were known to mount a cavalry charge against a similar enemy attack, to fight dismounted against other dismounted cavalry aided by mounted flank attacks, and to attack dismounted cavalry holding them off with firepower. In the West, Southern cavalry charged at the full gallop, firing both barrels of a double-barrelled shotgun in their opponents' faces, before drawing revolvers or laying about them with the butts of their guns.

Uniforms

Confederates

Regulation issue was similar to that of the Infantry – a short 'shell' jacket of grey with high collar, and light blue trousers worn inside or outside the boots. Kepi was yellow, with black peak and dark blue band around bottom. Jacket had yellow collar, cuffs and trim down front and around bottom. Trousers had yellow stripe. NCO's chevrons were yellow. A slouch hat often replaced the kepi as the war progressed, and was usually dark grey or black. Equipment was revolver, carbine and sabre; the carbine was carried slung from the shoulder belt, barrel down behind the right leg. The sabre was steel with gilt fittings. Field Officers wore slouch hat often with plume, or kepi and red waist sash. Scabbard black leather. Many favoured buff corduroy breeches or their pre-war Union blue trousers. Collar badges and gold braid on sleeve denoted rank.

The most famous of all Rebel Cavalry was Jeb Stuart's 1st Virginia Cavalry, who wore grey jacket and trousers with black

---

Colours and Battle Honours
In battles of the seventeenth and eighteenth centuries the seizure of regimental colours was a prime objective, often the size and importance of a victory was in proportion to the number of captured flags. A system of battle awards brings an added interest to wargames, for example the loss of a colour and the acquisition of a battle honour can be worked into one's morale rules so as to be a lasting result of victory or defeat. The following awards system is suggested for the wargamer who wishes to incorporate this method into his campaigning.

A victorious army
All units actually engaged will have the name of the battle inscribed on their regimental flag.

The victorious army will capture the regimental flag for each enemy unit captured or wiped out.
continued

slouch hat, with a yellow cord around base of crown. Jacket had black collar, cuffs and shoulder straps and frogging across front. They were known also as the Black Horse Cavalry from their mounts. Officers wore a black plume on their hats; the usual chicken-guts on sleeve and yellow sash.

The 8th Texas Cavalry (Terry's Texas Rangers) looked like Western Cowboys, wearing civilian clothes and Confederate or captured Union items where possible. They embellished their equipment and clothing with the silver Texan 'Lone Star' and favoured the slouch hat. The sabre was rarely carried, being replaced by a revolver. The Rangers carried a knife and two revolvers on their belts, and often two more on the saddle. Double barrelled shotguns originally carried were later replaced by carbines and rifles. The hat should be black with silver star at front; scarf was pink or red; shirt usually white, pink or checkered; jacket dark blue, brown or grey with a red stripe down outside seam. Belt had silver star buckle and stars decorated each holster.

Federals

Regulation was dark blue short coat with high collar, and kepi and sky blue trousers. Jacket was piped down front, round bottom, round cuff, all around collar and down back seams in yellow. NCO's chevrons yellow. Brass shoulder scales were worn and can be represented by painting. Trousers had yellow stripe. Kepi had crossed sabres on crown, the black slouch hat also worn had them at front. Equipment as for Confederates. Officers with shoulder bars.

The one Lancer regiment in the war was 6th Pennsylvania (Rush's Lancers). They wore regulation uniforms, without shoulder belts, and carried 9ft (3m) lance of Norway fir with 11in (28cm) steel tip. Make this with 40mm length of wire with scarlet swallow-tailed pennant 4.5mm from tip.

All troopers, Rebel and Yankee, had a second smaller roll in front of saddle. Saddle-cloths varied but were often dark grey or blue with red, black or yellow stripe.

The Artillery

The field artillery of the Civil War was light and mobile so that it could move with the army in the field and be freely manoeuvred in battle, although its tactical use was rarely fully comprehended by generals of either side, handicapped by many major actions being fought over broken country preventing best use of the artillery-arm. The basic unit of artillery was the battery, usually of six guns with a captain in command; two guns formed a section under a lieutenant, comprising two platoons under a sergeant (chief of piece) and two corporals. Each gun (or piece) was drawn by a six-horsed limber that also included an ammunition-chest, two more being carried on a six-horsed caisson accompanying each piece. The six-horse teams had three drivers riding the horses on the left side; a battery of six

---

A defeated army

All units which have performed exceptionally in the battle and the retreat will have the name of that battle inscribed on their regimental flag.

Any unit which shows signs of cowardice or panic in battle is not awarded any honours.

The small-scale colours carried by regiments of 20mm figures do not permit any very extensive lettering upon them – unless the wargamer has that sort of talent which permits him to inscribe the Lord's Prayer on a grain of rice! Give each regiment its colours and for each one paint up a larger replica, say about 4× 4in (10×10cm). These replicas can be mounted on a large base and displayed as a colourful wall decoration for the wargames room. Each colour will be suitably embellished with every fresh battle honour bravely won.

The Indian Mutiny and The Crimean War

The Indian Mutiny and the Crimean War were vastly different campaigns – in Russia the land operations were limited in area and scope; in India they extended over a greater part of the sub-continent. The Crimean War resembled the Napoleonic campaigns that ended at Waterloo whereas the Indian Mutiny was not a Victorian small war but a superiority of armament lay all on one side. Although lacking leadership, many of the enemy in India actually possessed better weapons than the British, while equipment and training were identical. The Indian Mutiny was a mutiny of the Bengal Army put down by British and Indian troops. The mutinous Bengal Army were dressed in scarlet coatees, white trousers, crossbelts and high shakos; their cavalry and artillery were dressed in the European fashion of the time. Invariably, they had superior numbers with plenty of guns; the mutineers had the climate in their favour and frequently they occupied fortified positions and buildings. However, if they were attacked in any way other than that in which they expected invariably they gave way without inflicting any great loss.

Every strategical and tactical principle of warfare seems to have been disregarded by the British commanders during the Indian Mutiny, so making it useless as a study of war. The inefficiency of the mutineers allows these errors and omissions to be perpetrated with perfect impunity. Nevertheless, by sheer dash and courage, the smaller numbers of British troops were able to impose their will upon the mutineers in *continued*

12-pounder guns, with two caissons per gun, required twenty teams with ten spare teams. A gun-crew was formed of nine men who rode on the ammunition-chest or walked beside the piece; horse (or flying) artillery gunners rode horses, with two men designated as horse-holders in action.

Confederate artillery were mostly in battalions, often formed by four batteries sometimes with only four guns each; each division had a battalion with two in reserve to back the five battalions attached to each corps. Union corps had artillery brigades formed of four or five batteries, with an Army reserve of twenty-four batteries; four batteries were attached to each division. The Union system of having divisional reserve artillery enabled them to effectively use massed batteries.

The gunners' role was a dangerous one, yet throughout the long war their courage and the manner of handling their weapons was first-class. To employ canister, their most effective weapon, gunners had to bring their pieces within 300yd (270m) of the target, whereas the increased efficiency of rifled-muskets meant that veteran sharpshooters were deadly up to 500yd (460m) range and could cause heavy casualties to gunners and horses.

The best known of all Civil War artillery was the 12 pdr Napoleon, the 1857 model smoothbore gun-howitzer; the most used rifled pieces were the 3in Ordnance and 10 pdr 3in Parrott, although imported British Armstrong and Whitworth breech-loaders were far more accurate than the muzzle-loading smooth-bore field-pieces. But their longer range – the Whitworth could throw its projectile nearly 6 miles (10km) – was of little advantage in this era of minimal fire-control and observation. These breech-loaders had relatively complicated breech-mechanisms with a separately-loaded charge and projectile, so that muzzle-loaders (which were far more used) could be loaded as fast.

When guns fired they emitted dense clouds of black powder-smoke, which made laying them difficult and often the direction of fire was only maintained by using the muddy tracks made by the wheels of the recoiling gun. There was no such thing as a recoil-mechanism, so guns hurtled back after firing and had to be run-out, re-aimed and pointed after firing each round; gun-sights were crude and inefficient so that aiming took more time than loading.

Both smoothbore and rifled guns fired solid-shot, shell, spherical-case (shrapnel), grapeshot, and canister – the last three usually being called case-shot. Solid shot was more accurate than shell or spherical-case and had a longer range; it was highly effective against troops in close formation. Shell was used against buildings and earthworks, and troops under cover with an effect more moral than physical as its bursting-charge was so small; the flash of the guns discharge ignited the time-fuse. It was the crudeness of the fuse that made it impractical to use spherical-case against rapidly moving targets, and it more often employed against bodies of troops at ranges from

An original photograph taken in 1862 of Confederate Volunteers at Pensacola in Florida

500–1,500yd (460–1,370m). Producing devastating effects upon massed troops, canister caused more casualties than all other projectiles put together; used at close ranges of about 300yd (270m) or less, it consisted of tin cylinders filled with iron-shot or musket-balls to give the effect of a monster sawn-off shotgun; in an emergency double canister with a single charge was used. Fuses were used to explode shell and spherical-caseshot but more often than not did not function correctly, so that the projectile either burst at the muzzle or not at all; they were percussion-fuses firing on impact or ignited by the flash of the guns discharge and timed to set-off the bursting-charge on or near the target.

Wargaming Aspects of the American Civil War

Perhaps because of its colour, diversity and magnitude the wars of the Napoleonic period are among the most popular tabletop conflicts; so much so that there is a marked tendency among wargamers to fight most pre-20th century wars along Napoleonic lines and many sets of rules are so compiled as to encourage this attitude. A close study of the American Civil War should speedily dispel this situation because it was this conflict that starkly revealed the future of land combat, although Europe, still nurturing the Napoleonic dream, saw it as a conflict of '... armed mobs chasing each other around the countryside, from which nothing could be learned.' However, its lessons were still appli-

almost every situation in which they came together.

During the later stages of the mutiny, in 1858, Major-General Sir Hugh Rose fought a masterly campaign in central India, displaying qualities of speed, activity and tactics up to then unknown in India. It was a campaign where khaki uniforms were first worn on a large scale as the troops marched more than a thousand miles in temperatures reaching 115°F (46°C) in the shade, capturing 100 guns, 2 fortresses, 2 cities, 20 forts and fighting 16 actions.

33

Favoured Formations for Wargaming – The Iron Brigade

On the morning of 1 July 1863, just west of Gettysburg in Pennsylvania, two brigades of Heth's Confederate Infantry Division moved forward somewhat carelessly against an enemy formation they thought were militia. Suddenly, the musketry-fire against them rose to a frightening and surprising crescendo and, while still reeling from it, out of the undergrowth onto their flank burst a determined line of bayonet-tipped attackers. Before being swept into rout, the startled Rebels just had time to shout: 'That ain't no Militia . . . look at them black hats . . . it's the pesky Iron Brigade!'

One of the most famous and best-known formations of the American Civil War, the Iron Brigade of the West was originally formed of the 19th Indiana and the 2nd, 6th and 7th Wisconsin Regiments; 2,400 men under command of General John Gibbon. Known as The Black Hat Brigade because of the non-regulation black slouch-hats they wore, the formation distinguished itself in its first engagement at Groveton in the 2nd Battle of Bull Run, when it sustained 33 percent casualties while fighting as part of King's Division against Stonewall Jackson's Corps. At this time, in three weeks, the Brigade fought five battles and lost 58 percent of its strength.

The 24th Michigan Regiment joined the Brigade just before Gibbon led them to further distinction in the Antietam Campaign, when a War Correspondent titled them 'The Iron Brigade'. Later, they fought at Fredericksburg and Chancellorsville, and then went *continued*

cable in World War I, because the tactics improvised by both Federal and Confederate soldiers developed into the methods of modern warfare. Lee's 7 Days' Campaign proved beyond doubt that Napoleonic warfare was dead and buried; the Confederate commander realised that tactical formations had to conform to the restrictions imposed by the accurate rifled-musket in the capable hands of first-class marksmen. Lee was the first leader to appreciate that the balance between offensive and defensive had been restored by the increased range of the rifled-musket and artillery, preventing close concentration of attackers. Gone were the short-range muskets and cannon of the Napoleonic era, that allowed the attacker to concentrate close to the enemy and win victories by successive blows of massed artillery, dense columns of bayonets and irresistible charges by heavy cavalry.

These factors *must* be reflected in rules if the American Civil War is to be realistically fought. For example, rules must reflect the essential differences and potentialities of the muzzle-loading rifled musket and the magazine repeater rifle and carbine, by giving the first two a long effective range, and all three a short-range longer than the short (cannister) range of artillery. Thus is exemplified the vulnerability of gun-crews to normal infantry weapons. Repeating rifles and carbines should be given a greater rate of fire, even twice per move – one shot at the beginning of the move and a second at its conclusion, with the repeater-carbine being given a shorter range. The muzzle-loading rifled musket should be given half a move to re-load, so that infantry can move half-distance and still fire.

The repeater-carbine in the hands of cavalry revolutionised the tactics of the mounted arm so that they did not have to rely on wild charges into the cannon's mouth, slashing away with the sabre in a hectic melee. Now they could dismount, out of the deadly canister-range of the guns, and from concealed and protected positions, pour volumes of fire into the unprotected gun-crews. This factor, coupled with the ability of normal line-infantry to similarly fire from a laying-down position, diminished the fearsome power of artillery and took away their title of Queen of the Battlefield. On a practical level, it also makes tabletop wargaming much more interesting and enjoyable!

Rules of Wargames in the American Civil War Period

The effects of being fired upon or taking part in a hand-to-hand melee are reflected by the *morale* of the unit *as a whole* being affected in that it is – *Unaffected*, when it remains where it is on the field, able to move, fire, or fight in the next game-move in normal style.

Shaken – when the unit withdraws its normal move-distance, still in good fighting-order, but not able to move forward (or towards the enemy) until the next game-move. It may fight in the normal way if attacked.

Disordered – when the unit withdraws its normal move-distance, but ends with its back to the enemy, unable to fight or fire until its owner is able to score 4 or over on the dice, which he may throw at the beginning of each subsequent game-move. Each failure to make the 4 means the unit retreats a further game-move distance. Whilst disordered, a unit can be fired upon or attacked without being able to retaliate and automatically moves back – in disorder – its move-distance.

Units do *not* suffer individual casualties and no figures are removed from the table. This means that the actual numerical strength of the tabletop unit is unimportant – it may consist of 20, 30, or more figures without being practically affected in any way, neither advantageously or detrimentally.

It is recommended that the figures making up a unit are grouped together in threes or fours on card bases.

A Union 12pdr Field gun in action

Moving

Before each move both commanders throw a die; the highest scorer has the choice of –

1. Moving FIRST and firing SECOND (he moves; his opponent moves and fires; he fires – all targets as they stand at the end of both moving)

OR

2. Moving SECOND and firing FIRST (opponent moves; he moves and fires; opponent moves).

Move Distances

Infantry Units (as a body)	6in (15cm) per move
Infantry when skirmishing	9in (23cm)
(denoted by groups being spaced)	
Cavalry	12in (30cm)
Horse Artillery	12in (30cm)*
Field Artillery	6in (15cm)

*A Horse-Gun takes 2in (5cm) off its move to limber up *or* unlimber.

All Arms – deduct ⅓ move for crossing walls, hedges; moving through woods; climbing hills. Add ⅓ move when on roads.

Firing

When both sides have completed their moves, each fires their guns and rifles, in that order, in accordance with the Move/Fire sequence described under 'Moving'. Targets are as they are at the time of firing, except when one unit attacks another (when it takes fire in its approach) and can be thrown back before reaching the unit it wishes to melee.

Firing Ranges

	Normal Range	Short Range
Infantry rifle or musket	18in (45cm)	9in (23cm)
Field Gun	48in (122cm)	6in (15cm)
Horse Artillery	36in (91cm)	6in (15cm)

into the Gettysburg Campaign as the 1st Brigade of the 1st Division of the 1st Corps; here they were commanded by General Solomon Meredith as Gibbon had taken over 2nd Corps, to be severely wounded at Gettysburg. In the severe fighting for Seminary Ridge on the first day at Gettysburg, the Brigade lost two-thirds of its 1,800 effectives, casualties in the 24th Michigan Regiment were 399 out of 496. Total Brigade losses were 162 killed and 724 wounded, 267 missing – a total of 1,153. From then on, although the Iron Brigade continued to function as a formation, it never recaptured its original qualities.

Wargamers invariably seek formations wearing 'different' uniforms, not unnaturally tiring of painting-up unit after unit in the same dress. Thus, when assembling armies for the American Civil War, such colourful units as those wearing Zouave-style uniforms (Ellsworth's Fire Zouaves, the Louisiana Zouaves etc) are favoured. This causes a quite anachronistic imbalance of forces, with a totally unrealistic *continued*

A Civil War battle as portrayed in a magazine of the day

proportion of Zouaves and the like, who perform well outside historical abilities although embellishing the tabletop battlefield.

The Iron Brigade are a different matter – they really existed as a first-class fighting formation who earned all the laurels given them and are more than worthy of adequate representation on the wargames table. Admittedly, they are not ALL that much different to the usual light blue pants, dark blue tunics of the run-of-the-mill *continued*

Simulation of Effects of Musketry or Artillery Fire

This chart shows the FIRER on the lefthand side; the TYPE of TARGET, in its specific state, is shown along the top. All scores are at normal range; at short-range, take the score on the extreme right.

FIRERS			TARGETS			Guns		
	In Open	Skirmishers	In Hard Cover	In Soft Cover	Cavalry	In Open	In Hard Cover	At Short Range
Rifles	2	1	0	1	2	2	0	3
Horse Gun	3	2	1	2	3	3	1	4
Field Gun	4	3	2	3	4	4	2	4

To use, say infantry in the open are being fired upon by rifles – on the line RIFLES look right and under column-heading 'In Open' is figure 2 which is the operative basic figure. For Cavalry under fire from a Field Gun – on 'Field Gun' line, look right to 'Cavalry' column; the figure 4 is the operative basic figure.

Simulation of Effects of Fighting Hand-to-Hand Melee

This chart shows the ATTACKER on the extreme left, and the DEFENDERS, in their specific state, are shown along the top. Each column is divided into two further columns – A (Attacker) and D (Defender) – the Attacker being the unit making the actual contact, the Defender receiving it; if two units contact simultaneously, then use scores shown in Attacker column for both.

ATTACKERS	DEFENDERS												
	In Open		Skirmishers		In Hard Cover		In Soft Cover		Cavalry		Guns In Open		In Hard Cover
	A	D	A	D	A	D	A	D	A	D	A	D	A	D
Infantry	3	3	2	4	4	2	4	2	4	2	2	4	4	2
Skirmishers	4	2	3	3	4	2	4	2	4	2	3	3	4	2
Cavalry	3	3	2	4	4	2	4	2	3	3	2	4	4	2

How To Use Firing and Melee Charts

Firing – target-unit throws a 1–6 die for each unit or gun firing upon it; then checks the chart to ascertain its operative base-figure, to which is added the score thrown upon the die. A score of 6 or over is a HIT. Thus, if base-figure is 2 and die score is 3, then it is a MISS; but if base-figure is 2 and die score is 5, then it is a HIT.

Melee – Both units involved in melee throw a die, adding its score to the operative base-figure from the chart. If 6 or over is totalled, then the unit concerned is SHAKEN.

Units HIT or SHAKEN throw die again –

Score of 1 = the unit is in DISORDER and immediately moves directly back its move-distance, ending with its back to the enemy. Read first paragraph of these rules for further action.

Score of 2 = the unit is SHAKEN again. Read first paragraph of these rules for further action.

Scores of 3–6 = Unit is unaffected and may carry on unimpaired.

NOTE: HARD COVER means being behind a wall, in rocks, or in a building.

SOFT COVER means being behind a hedge, or in a wood.

When a DISORDERED unit has not been 'recovered' by throwing 4 or over on die and reaches its baseline, it is removed from the table – and the game.

Federal formation, but they do have different hats to mark them out as the favoured formation they are. Those whose ACW armies are formed of the economical (if obtainable) Airfix H0/00 scale figures will find the ideal model among the selection, wearing a black slouch hat (albeit turned-up at the side a bit like a Napoleonic Austrian Jager) but presentable.

The Iron Brigade should be granted *elite* status in wargames operations, in the same manner as Guard-units in wars of other periods; given higher morale-status and perhaps enhanced fighting ability. They will stand-out, both in appearance and performance!

3
'THEY CAME IN LIKE GREAT BIRDS . . .!'
The Storming of Eben Emael, 10 May 1940

Nothing can approach actually walking round an historic battle-field, reasoning how it was won or lost, to inspire a wargamer to re-fight it on a tabletop terrain; studying the shell-pocked entrance and picnicking on the grassy ridge hiding the tunnels and gun emplacements of the World War II Belgian fort of Eben Emael made this a battle that *had* to be re-fought. Wargames tend to conform to a pattern but this one was really different, the stirring events of the 10 May 1940 provided all the ingredients required for a satisfying tabletop encounter – colourful military and historical background to provide interest and competitiveness; large enough to be worthwhile yet well within the reach of even the schoolboy's pocket; and, even if the Baddies did beat the Goodies, it was an astonishing example of a triumph over seemingly insurmountable odds through the use of unique technological factors. The very first of its type and virtually the beginnings of a War, this action transformed airborne warfare from a theory to a fact; never before had there been an airborne operation under actual combat conditions against really determined opposition.

An essential ingredient in any German breakthrough to the Channel Coast in 1940 was an attack through Belgium by a rapid break-through of Belgian covering positions on the Albert Canal close to the Dutch border, only 15 miles (24km) from Germany itself. In the Spring of 1940, although hoping to maintain strict neutrality and conscious that Britain and France would aid if she was attacked, Belgium relied on a delaying defensive position protected by a forward line of outposts. However at Maastricht, where the nearness of the Albert Canal to the Dutch border made outposts impossible, reliance was placed upon the Albert Canal's deep cutting, its 100yd (90m) width being spanned by three bridges at Veldvezelt, Vroenhofn and Canne. The 17th Infantry Division were responsible for defending the area and had a brigade covering each bridge. The whole position was supported by the powerful artillery fort of Eben Emael.

Well prepared and sited, the bridge defences consisted of four massive concrete pillboxes on the near bank, one beside the road mounted an anti-tank gun; another immediately below the bridge, and one on each flank some 500yd (460m) distant, all mounted machine-guns. There was also a small post on the far (eastern) bank. The positions were garrisoned by a company positioned on the near bank at each bridge. Canne, the southern-

Fort Eben Emael, Belgium. On top of the fort, which is within the massive grassy and wooded mound on which this photo was taken in 1987 by D. Featherstone

Fort Eben Emael, Albert Canal, Belgium. Stormed by German Assault glider troops on 10 May 1940. A ground-level gun emplacement; although it is difficult to detect, the battered gun is still in position

continued

The Use of Armour Between the Wars, 1918–1939

The armoured car policing of Palestine

Armoured Cars on the North-West Frontier of India

Vickers light tanks on North-West Frontier

The Italian invasion of Abyssinia

Armour in the Spanish Civil War

Tanks in the Russo-Japanese conflict of 1938–9

Wargaming Aspects:

The inter-war years can be a fascinating period for the wargamer seeking something different, who wishes to re-fight tank actions without being bogged-down by the complex technologies of armoured war in World War II and since.

The wargamer fighting Colonial Wars, by simply including a single armoured car, will completely revolutionise the style of warfare, and the 'Native commander' will be forced to discover elementary or primitive methods of combatting the threatening vehicle, just as did his counterpart in real life.

A motley collection of armoured vehicles were employed in the Spanish Civil War 1936–9 and the wargamer seeking something different (particularly if his interests extend to scratch-building vehicles) will discover much of interest in this, the largest of the inter-wars conflicts.

For more than a hundred years the North West Frontier of India was a constantly running sore in the side of British rule. Nevertheless, it provided an ideal training ground for British soldiers to learn to take cover correctly, to patrol, to fire accurately and in a controlled manner, and gather the basic

most of the bridges and nearest to Eben Emael, had an anti-tank gun bunker set back into the hillside.

All the bridges had prepared demolition charges in position and could be quickly blown by demolition parties in the anti-tank bunkers. No Belgian forces operated east of the canal because of the nearness of the Dutch frontier but surprise seemed impossible because the Germans would have to fight their way across the 'Maastricht appendix' of Holland, so that by the time they reached the Belgium border the bridges would be demolished and well prepared defences confidently awaiting them.

With bitter memories of 1914 when the forts around Liège had been smashed into submission by heavy German siege-guns, in 1933–5 the Belgians had blasted out of natural rock the fortress of Eben Emael resembling the great defensive works of the French Maginot Line, with one side rising a sheer 120ft (36m) from the canal. The other faces were protected by concrete pillboxes, 60mm anti-tank guns, heavy and light machine-guns, ditches, a 20ft (6m) wall, mine-fields and searchlights. The fort's armament consisted of six 120mm guns in revolving armoured cupolas and eighteen 75mm guns in cupolas or casemates, mounted in emplacements with walls and roofs of 5ft (1.5m) thick reinforced concrete.

In November 1939, a special combat group under Hauptmann Walter Koch was formed from 1st Parachute Regiment 7th Air Division of Engineers, plus pilots, for the special task of capturing the three bridges intact and neutralising the fort. STURM ABTEILUNG KOCH were all paratroopers, nevertheless it was decided the attack should be made in gliders, Hitler having been told by Hanna Heitsch, a celebrated sports glider-pilot, that a glider in flight was almost silent. For six months Koch's forces trained extensively on full-size mock-ups of the Eban Emael defence system until every man was well-versed in his own role as that of his comrades in ensuring a concentration of gliders on the various objectives. Using the DFS230, the first military assault glider, they were to be towed to the Dutch border and released at 8,000ft (2,400m), to glide silently across the Maastricht 'Appendix' and onto the bridges and the fort, undetected by sound-location devices of Belgian anti-aircraft defences.

The DFS230 was developed as a means of delivering an assault from the air without separating the troops from their weapons; it could carry heavier armament directly into battle. This enabled the German paratroopers to be safely landed fully equipped with MP40 sub-machine guns and MG34s ready for action, obviating the need to find a weapons' cannister. This glider possessed a defensive armament of a single 7.92mm machine-gun 15 mounted on the upper deck of the forward fuselage, with a flexible mount allowing it to give supporting fire to the occupants of the glider; it could be removed and taken into the assault. Specification: Weight loaded – 4,600lb (2,100kg); Towing Speed – 131mph (210kph); maximum speed – 181mph

(290kph). Each carried a pilot and co-pilot, sitting side-by-side, the first pilot being on the left; also nine fully equipped men who could exit rapidly through doors at each end. The controls were similar to those of a power-driven aircraft of the day, except that there were no throttles; instead, a small lever (painted red) operated the tow-rope release. On operations, a JU52 towed one glider, three JU52s, with their gliders, flying in formation.

The plan was that just before 0530 hours, when the main offensive was due to begin, gliders would silently land on the western bank beside each bridge (that is, on the Belgian side of the Albert Canal). Koch's assault force was in four distinct groups, each with specific duties: (1) Assault group 'CONCRETE' (Leutnant Schacht) to secure the bridge over the Albert Canal at Vroenhofen and hold until the arrival of ground forces; (2) Assault Group 'IRON' (Leutnant Schaechter) to secure the bridge at Canne; (3) Assault Group 'STEEL' (Oberleutnant Altman) to secure the bridge at Veldvezelt. Each glider would contain about 90 men formed of 5 Infantry and 4 Engineering Sections, who would surprise and overwhelm the defenders, remove prepared charges under the bridges and then prepare to defend the area against the expected counter-attack. Forty minutes later, three JU52s would fly over each of the objectives, parachuting-in 24 reinforcements, machine-guns and ammunition. Meanwhile, Assault Group GRANITE (90 men all Engineers) under Oberleutnant Witzig – in 11 gliders – were to land on the flat roof of the fort, fight-off defenders and then set about crippling the artillery armament, while preventing the garrison from dislodging them from the roof. These trained and experienced

facts of battlefield survival. Wily tribesmen, the enemy knew their country like the back of their hands, so that they could blend into the rocky landscape and disappear or reappear at will. Crack shots, they frequently used stolen British weapons. In the latter years of this long-standing duel, armoured cars and light tanks played vital roles, but the hillmen utilised their knowledge of the terrain and ingenuity in devising weapons, so that they continued to hold their own.

In 1921 the Indian Government began to open up territory by garrisoning tribal areas; roads were built in each area and a section of four armoured cars patrolled a given stretch. The cars were invaluable to cover the withdrawal of punitive expeditions, by keeping the sniping tribesmen at a distance; and elsewhere, by a show of force, the armoured cars kept troublemakers quiet and prevented racial and inter-sectarian disputes from flaring up in open disorder.

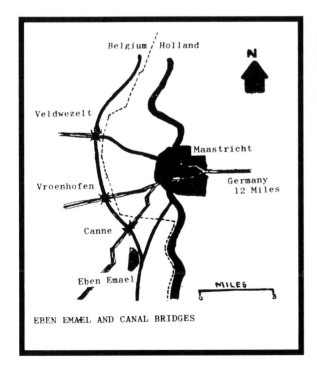

EBEN EMAEL AND CANAL BRIDGES

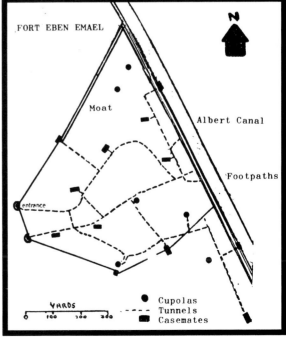

The Canne Bridge over the Albert Canal alongside Fort Eben Emael on the border of Belgium and Germany. This bridge was blown on 10 May 1940 before the assaulting German Airborne troops could get over it

Plan a Wargame No. 4

Two equal strength armies (in points) are decided upon and then divided into three groups, each to the choice of the commander. The first group consists of 25 percent of the Army's points total – this is the Advance Guard. The second group, the Main Body, totals 50 percent of the points and the third group, the Rear Guard, totals 25 percent.

Erect a curtain across a perfectly clear wargames table, and then divide some pieces of terrain equally between both commanders, who will set up the battlefield on their side of the curtain. Thus, neither commander knows the ground towards which he is advancing as it is assumed that there is a heavy mist.

Both sides now lay their Advance Guard out in any desired dispositions on their own side of the curtain. When this is completed, the curtain is raised and the battle begins in the normal way.

As soon as the curtain is lifted, each commander *privately* throws a dice to decide whether he is an Exceptional Commander (5 or 6); an Average Commander (4 or 3) or a Below-Average Commander (1 or 2) – neither commander is aware of his rival's rating. As explained in Chapter 1, this affects everything a Commander does in the battle, including his reaction to this sudden encounter with the enemy. Dependent upon his *continued*

engineers were armed with 110lb (50kg) hollow-charge explosives capable of punching a hole 12in (30cm) in diameter through 6ft (2m) of concrete.

The flight-path called for precise navigation as it would be in darkness, ground-beacons marking the route to the frontier; innumerable rehearsals perfected techniques of casting-off twenty miles from the objectives and gliding in to a pin-point landing at daybreak. At 0415 hours on the 10 May 1940, forty-one JU52s, each towing a glider, began taking-off and wheeled into formation to steadily climb for thirty minutes to just over 8,000ft (2,400m) in a much-practised technique. The correct position to be assumed by the glider is either slightly above the tug, high-tow, or slightly below it, low-tow. The glider must never be directly behind the tug, for the slipstream would cause it to oscillate so violently that the tow rope would soon snap. The breaking of the tow rope is naturally the chief fear of the pilot, and it may assume the proportions of a nightmare if he is flying in cloudy and bumpy weather. The strength of the rope depends on the strength of the tug's fuselage and the nose of the glider, to which it is attached by a simple bolt-and-shackle device. Too strong a rope would mean putting too great a strain on the tug; too weak would lead to frequent snapping. In fact, one rope did snap and of all people it had to be Leutnant Witzig commanding the Eben Emael group, who landed in a field deep in Germany.

In each glider shivering men pulled blankets closer round shoulders and knees; the passage of the craft was silent to those on the ground, but the wind-noise is considerable even in free-flight; in-tow it makes conversation difficult, although the assault-parties had too much to occupy their minds to be bothered about talk. Finally, the formation reached the furthest point at which towing-aircraft could turn back without crossing the frontier, the tow-ropes were let go and the silent gliders were in free flight, less that of Leutnant Witzig, and one other which,

The Eben Emael Wargame. A German DFS 230 Assault glider lands alongside a gun-cupola and its crew go into action

casting-off too soon, never reached the objective. They nosed gracefully down towards the Belgian frontier, getting lower and lower until one by one their skids touched down and they ran forward twenty yards before stopping.

On 10 May, even before a wing tip touched ground doors were off and soldiers poured out, as each bridge group (consisting of 5 Infantry and 4 Engineer sections) ran swiftly towards the pillboxes and bridge, returning fire opened upon them from the bridge defences. The advanced proceeded in leaps and bounds until the Engineers were within striking distance of the bunkers, dramatically neutralised by having huge holes blown in their concrete by hollow-charges and then pouring flame and throwing grenades through the gaps. The two northern bridges at Veldvezelt and Vroenhofen were seized intact and demolition charges removed before they could be blown. The Belgian demolition firing party at Veldvezelt realised that the bridge was going to be captured and radioed to their HQ, 3 miles (5km) north, for permission to blow but, disbelieving the story of a glider attack and the presence of German troops on the bridge, permission was refused.

At Canne, surrounding hills delayed the gliders and they put down several hundred yards from their objective, so forfeiting surprise. The defenders put up a heavy fire that prevented the engineers storming the pillboxes; so permission for it to be blown was readily given, and it went up in the face of its frustrated attackers.

At Eben Emael, the nine remaining gliders of Witzig's group landed with precision on the roof and disgorged attackers who ran at will all over the exterior of the fort systematically destroying the twelve emplacements from which fire could be brought to bear on the bridges and the surfaces of the forts. The paratroopers fired and flamed, threw grenades into the embrasures and loopholes, placed charges of TNT on turret-edges and gun

rating, he will at some stage send back a courier to warn the Main Body. At the same time each General dices to ascertain the rating of the commanders of his Main Body and his Rear Guard.

The time at which this courier leaves, and the time he takes on his journey may be estimated either by using actual clock-time or else by using the number of game-moves. Thus, he may leave during the first game-move and arrive at the Main Body on the third game-move. The gap between the Advance Guard and the Main Body (also the gap between them and the Rear Guard) must be decided in the same way.

When the Courier leaves, each commander will draw a 'Courier Card' (see page 117). This card will *not* be looked at until enough time has elapsed for the courier to reach the Main Body by the swiftest means (he may well have drawn a Courier Card that causes him to delay en route).

When the Courier reaches the Main Body, the Commander there acting in accordance with his status, will assess his reactions to the news and he will act accordingly in moving forward or otherwise.

Whilst this has been going on, the heavy mist that earlier covered the field has lifted. A 'Weather Card' is now drawn to discover whether it will affect speed of movement etc (see page 137).

And so, the battle proceeds with each Advance Guard Commander endeavouring to take and hold the most advantageous positions for his *continued*

Main Body on their arrival. At the same time, each commander is anxiously looking back for his own Main Body and praying that the enemy's Main Body do not arrive first. This, of course, is dependent upon his initial reaction to the sight of the enemy; the speed at which his courier got back to the Main Body and the sense of urgency displayed by its commander.

As each Main Body (and Rear Guard) arrives, they are fed into the battle.

This is an ideal method of wargaming where a number of players are available – each player can take a command. But, the most valuable use for an extra man is for him to act as an umpire, handling all the cards and keeping each commander informed as to times etc.

barrels, jamming the turrets and destroying the guns; ventilators and periscopes were attacked. Hollow charges detonated on top of the emplacements and turrets blew holes through the armour and concrete, sending a jet of flame and molten metal into the turret structure to wreck internal machinery and kill or shock defenders. Those who survived were assailed by flame-throwers or small charges and grenades dropped through the holes. Blasting open steel doors, the invading paratroopers entered the fort and, once inside, were difficult to eject as the defenders had to attack up 60ft (18m) of spiral staircase.

Crossing the Albert Canal in inflatable boats, pioneers of an Engineer battalion brought heavy demolition charges, flame throwers, a Bangalore torpedo for wire cutting, and other materials for prising open the fort.

At 0830 a lone glider flew in from the East, across the Dutch and Belgian frontiers, circled over Eben Emael and touched down in one of the few remaining clear areas – it was Witzig, towed off by a relief-aircraft, arriving three hours late to take part in the assault.

At 0610 hours German aircraft, taking casualties from now fully alerted defences, dropped reinforcements and ammunition at each of the bridges. Weak Belgian counter-attacks against the captured bridges were repulsed, and the Canne Bridge group, strengthened and reorganised, cleared the last defenders from the demolished structure.

In a few hours, at a cost of 6 dead and 20 wounded, the 70 paratroopers had neutralised Eben Emael. The successful German airborne assault had come to be regarded as a classic example of what can be achieved by a small force against powerful defences, exemplifying the fullest exploitation of all the advantages that lie with the attacker. Tactical surprise was achieved by the silent approach of the gliders in a never-before-demonstrated landing method against which no defensive tactics had been considered.

Stirring stuff, isn't it?

By now even the most stolid wargamer should be rarin' to go; although an unusual game to play, it is not difficult to re-create and can be handled by scaling down forces 1:2 so that one model soldier represents two men in real life, handled by buying four boxes of HO scale plastic figures, currently selling at £1.75p a box. The beauty of this game lays in the fact that only one side has to be provided, as the Belgian defenders do not appear on the table, being represented by drawing from an ordinary pack of playing cards. On the day there were forty-one gliders, scaled-down to twenty-one for our game – these are simple flat affairs, made by tracing or photo-copying the plan of the glider in the illustration and colouring it a mottled green. The tabletop terrain is easy to set up, being quite flat except for some high ground in the bottom right-hand corner – the grassy mound that is the fort of Eben Emael. Each glider must be numbered from 1 to 21 and listed on four GLIDER CHARTS,

The inexpensive plastic HO/OO scale troops used in the Eben Emael wargame

Veldeweltz Bridge		Vroenhofen Bridge		Canne Bridge		Eben Emael	
SPADES		CLUBS		HEARTS		DIAMONDS	
1 King	2 Queen	3 King	4 Queen	5 King	6 Obstacle Queen	7 King	8 Queen
9 Obstacle Jack	10 Ten	11 Jack	12 Obstacle Ten	13 Jack	14 Ten	15 Obstacle Jack	16 Ten
17 Nine	18 Obstacle Eight	19 Nine	20 Eight	21 Obstacle Nine	22 Eight	23 Nine	24 Obstacle Eight
25 Seven RIVER Veldeweltz	26 Six RIVER Bridge	27 Seven	28 Six Obstacle	29 Seven	30 Six RIVER Canne	31 Seven RIVER Bridge	32 Six RIVER
33 Five ―DEFENCES―	34 Four	35 RIVER Vroenhofen Bridge Five	36 RIVER Four	37 RIVER Five	38 Four ―DEFENCES―	39 Five FORT	40 Four FORT
41 Three	42 Two	43 Three ――DEFENCES	44 Two	45 Three	46 Two	47 FORT Three	48 FORT Two

Altman's STEEL Command	Schacht's CONCRETE	Schaechter's IRON	Witzig's GRANITE
Glider 1	Glider 6	Glider 11	Glider 16
2	7	12	17
3	8	13	18
4	9	14	19
5	10	15	20
			21

LANDING-ZONE AND GLIDER CHARTS.

This is the chart by which the game is played – without it there can be no game!

one for each German commander, on which he lists against the appropriate glider what has happened to it in flight or on landing. A LANDING-ZONE CHART is required, scaled to the wargames table in 12in (30cm) squares (for a table 8×6ft (2.4×1.8m)), each square representing one of the designated landing-zones *or* a hazard that could, at worst, destroy both glider and its occupants. The landing-zone chart (which is really a gridded map of the terrain) is divided into four longitudinal sections, in three of them the bridge-assaults take place and in the fourth the attack on the fort; each section is code-named according to a suit of the playing card pack, thus:–

Veldwezelt Bridge	=	SPADES
Vroenhofen Bridge	=	CLUBS
Canne Bridge	=	HEARTS
Eben Emael Fort	=	DIAMONDS

Four wargamers represent the four commanders of the attacking groups; they allocate gliders – five each to the bridge-assault groups, and six to the Eben Emael group, as follows:–

1. Oberleutnant Altman's STEEL Command attacking Veld-wezelt Bridge, operating in SPADES squares has Gliders Nos. 1–5.
2. Leutnant Schacht's CONCRETE force attacking Vroenhofen Bridge, in CLUBS squares has Gliders Nos. 6–10.
3. Leutnant Schaechter's force IRON attacking Canne Bridge, operating in HEARTS squares, has Gliders Nos. 11–15.
4. Leutnant Witzig's GRANITE force at Eban Emael, operating in DIAMONDS squares, has Gliders Nos. 16–21.

All Bridge Assault groups are formed of three infantry sections and two Engineer Sections; the Eben Emael group are all Engineers. Gliders must be designated as carrying either Infantry or Engineers, because their different roles play a big part in the operations.

As Leutnant Witzig would ruefully testify, things can go wrong in flight; this is simulated in our reconstruction by drawing from an ordinary pack of playing cards–

Ace of Spades or Hearts = pilots casts off too soon and glider drops short of battle area.
Ace of Clubs or Diamonds = Tow rope breaks, with same effect as above.
King of Clubs or Hearts = Structural weakness causes wing or tail to drop off and glider plunges to ground, killing everyone.
King of Spades or Diamonds = Faulty take-off; glider crashes and takes no part in operations.

The game begins by each German commander drawing one playing card from the twelve of his personal suit, and placing

---

The Thirty Years' War
The Thirty Years' War was largely fought out with conventional arms but during the 17th century two relatively new weapons were developed on the Continent. One of them was the hand-grenade, a hollow cast-iron globe a few inches in diameter containing 2–3oz (55–85g) of powder and with a short fuse which was lit before throwing. A new type of soldier, the Grenadier, arose to use this weapon and in 1667, four or five Grenadiers were attached to each company of French infantry and three years later separate companies of Grenadiers were formed. Their equipment included a fusil or light flint-lock musket and a pouch of grenades with a slow match in a case. Grenadiers were usually tall strong men with long limbs who could throw their missile the greatest distance. A British Grenadier of the period wore buff cross-belts over a red coat and a waist belt with a frog at the left front; he wore a furred cap with a baggy crown which hung down the back, later to be stiffened upright to form the mitre cap of the 18th century soldier, worn by Grenadiers and Fusiliers. The second innovation was a close-quarters arm, the bayonet. Initially it was a dagger stuck by its tapered grip into the musket muzzle to form a complete offensive and defensive unit for one man. It was a development that out-dated the old pike and musket combination. It was not an easy weapon to use because when the socket of the bayonet was plugged into the musket, it was not possible to fire the weapon.

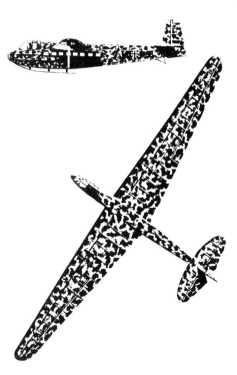

From this diagram the tracings are taken to make the gliders used in the game

Minefields

The minefield was not the imagined unmarked and seemingly innocuous area of ground thickly sown with deadly explosive charges waiting to destroy the unwary person or vehicle who unwittingly ventured into its confines. In a defensive context, minefields in areas of operations were in-situ long before the particular action, their siting known to the defenders who laid them. Their purpose to slow down enemy attacks and allow time for defenders to concentrate; or to 'channel' the enemy onto a 'killing-ground' registered by artillery and covered by fixed-line machine-guns. Often laid obviously, the psychological fear they inspired could cause the enemy to move his forces into such fire-covered *continued*

his first glider in the square indicated by the card. For example, Oberleutnant Altman draws three of Spades and his first glider lands in the bottom lefthand square of his operational area; he replaces the card then draws for his second glider, which lands in the square marked Jack of Spades – the last square-but-one at the top of the table, and so on until all his five gliders have landed (subject to accidents in flight). Each time a card is drawn it has to be replaced before a card for the second glider is drawn; record against the glider's number on the chart, indicating whether or not it has reached the zone of operations and where it has subsequently landed. As soon as the glider's landing-area is known, it can be placed upon the table in the appropriate square – these are not actually marked on the tabletop, but it is easy to measure from the edge to discover the centre of a square.

Off the table each Commander has his force of forty-five paratroopers and ten pilots (who fight as infantry) and, when all gliders have been placed in their squares, the soldiers from that glider are placed in its square and become operational when the next phase of the game begins. Should the glider have landed in a square containing an obstacle it will have suffered damage and casualties among its crew, revealed by throwing *two* ordinary dice (1 to 6), one white and one coloured, the former indicating the degree of damage suffered by the glider, thus:

Score of 1 = the glider and all its occupants are totally destroyed.

Score of 2 = the glider is damaged. The score on the coloured die indicates how many of its occupants are killed and therefore do not get onto the table.

The gliders have landed, the assault-troops are ready to move towards the three bridges and the fort, concealed under its great grass- and tree-covered mound, looms out of the grey light of dawn. The defenders are still reeling under the shock and surprise of seeing this fleet of silent aircraft 'coming in like Great Birds'; some are already fighting desperately against enemy who have landed in the very midst of their positions. Knowing their first duty is to deny the bridge to the enemy, the Belgian commanders are frantically telephoning Headquarters asking permission to blow the bridges. The delay in making a decision, or the fact that attackers have got onto a bridge and cut the charge-wires are factors simulated by *two* dice being thrown at the start of each move – if *both* dice reveal the same score ie – two 1s or *two* of any number, then the bridge goes up; should this not have happened by the time *four* attackers (who must be Engineers) actually get onto the bridge, then they can remove charges or cut wires in *two* moves, during which the defenders carry on trying to blow the bridge.

Move One – bridge-blowing dice have been thrown; the attackers move forward 6in (15cm) per move, but come under fire and encounter other problems, simulated in our game by drawing playing cards. If each group has its own pack of cards it will obviate waiting for a turn and while each commander is drawing his own cards he does not have time to see what is going on around him and reacting. There are *seven* 'threats' to the attackers – Rifles; Machine-guns; Grenades; Mortars; Artillery; Barbed Wire; Minefields, represented by suits of cards and/or individual cards of that suit. Their effect is divided into MAXIMUM, AVERAGE and MINIMUM, reflecting good/bad aiming; target under cover; etc, etc, and LUCK. Each 'Threat' bears a points value, deducted from the total number of points granted to a force. This is 3 points per man; a full-strength force of 45 paratroopers and 10 pilots = 165 points. Each commander has a Chart bearing numbers from 1 to 165, and when his group comes under fire or encounters an obstacle that costs them points, that number of points is marked-off on the Chart; and a soldier is removed from the table for each *three* points deducted. When a group takes MAXIMUM fire, all go to ground and do not move; AVERAGE fire causes them to move next move at half-rate; MINIMUM fire merely deducts specified points.

The number of 'Threats' encountered each move is decided by throwing a die, the score denoting the number of cards drawn from the pack indicating what attackers are encountering. Thus, die score of 3 – that number of cards are drawn – say, Jack of Spades; 2 of Diamonds; and 9 of Hearts. The first represents Maximum rifle-fire and costs 3 points; the second is Minimum Grenade damage and costs 3 points; the last is Average artillery-fire and costs 5 points, totalling 11 points which removes three men from the table (with 2 points towards fresh casualties next move); it also means that the group have got to go-to-ground, and – next move – they can only advance at half-rate. This is the table that controls the game:–

positions. Of course this happened when minefields lost their markers, or the darkness and stress of battle might cause panicky units to blunder into a field – but that could occur also to the unit that laid the mines! Generally, minefields were marked by low wire fences or perimeter notice boards bearing warnings in English or German, as the case might be.

Although minefield concealment has been explained, no doubt the wargamer will wish to make attempts to hide these deadly weapons, and it can be simulated in a realistic manner with thin coloured transparent plastic sheeting. Commanders are given a specified area of this material, say 8×8in (20×20cm), from which to cut as many irregular minefield patterns as desired. Laid on similar coloured terrain, these coloured transparencies are difficult to identify without close scrutiny and are sufficiently unobtrusive to allow occasional blunderings onto them.

There are two types of mines – anti-tank and anti-personnel. The effects of the former upon vehicles can be simulated by ruling that the tank has a 25 percent chance of getting through and a soft-vehicle a 10 percent chance – decided by percentage dice thrown each move that the vehicle is in the minefield. The effects of anti-personnel mines are represented by ruling that a man has a 20 percent chance of escaping, percentage-dice being thrown each game-move if the man is actually moving – as long as he stays still he is safe. On the wargames-table, mines can be laid – per game-move – over an area of 3×3in (7.5×7.5cm) by Engineers or other troops who cannot fight or take any other action while so doing.

Weapon	Cards	MAXIMUM	Points	AVERAGE	Points	MINIMUM	Points
Rifles	Spades	Ace King Queen Jack	3	7 8 9 10	2	2 3 4 5 6	1
Machine-Guns	Clubs	Ace King	5	9 10	4	5 6	3
Grenades	Diamonds	Ace King Queen Jack	5	7 8 9 10	4	2 3 4 5 6	3
Mortars	Clubs	Queen Jack	6	7 8	5	2 3 4	4
Artillery	Hearts	Ace King	6	9 10	5	4 5 6	4
Barbed wire	Hearts	Queen	3	8	2	3	1
Minefield	Hearts	Jack	5	7	4	2	3

Despite a card being drawn it does not always apply if enemy are not within its 'killing-zone', thus:–

Helicopters in Action

Glider-borne landings have not been performed since World War II; and para-drops have been used only in small-scale operations, such as in 1978 when the 2nd REP of the Foreign Legion were dropped into Kolwezi in Southern Zaire to rescue 3,000 European residents from Katangan 'Tiger' rebels. All modern infantry take on the airborne role, transhipped to the area of operations by troop-carrying helicopters, the airborne equivalent of the armoured personnel-carrier, but less susceptible to terrain difficulties and unlikely to be knocked-out during the approach-march.

The sudden arrival on the wargames table of helicopter-borne infantry can be simulated as para-drops or glider-landings; or troops can be deployed on the table, their airborne drop being assumed to have taken place. Scaled-down in size and numbers, model helicopters will add colour and stimulus to a wargame; their use governed in much the same way as suggested for gliders, but with far greater degrees of accuracy in landing-zones.

Helicopters can also play the role of mobile artillery, capable of bringing to bear immense fire-power in anti-personnel and anti-tank roles, through their varied armament of heavy machine-guns, rapid firing cannon, rockets and other modern missiles.

Against Bridge-Attackers

Rifles; Machine-Guns; Barbed-wire – effective anywhere on table.
Mortars; Artillery; Minefields – effective only in area beyond Canal.
Grenades – effective only in defence area on Belgian side of canal (including bridge area).

Against Eben Emael Fort Attackers

All weapons except Grenades are effective in Squares (Diamonds) King; Queen; Jack; 10 9 8 7 6.
Mortars and Artillery are NOT effective *on the actual fort*, ie Squares (Diamonds) 2 3 4 5.

Finally, one further penance inflicted upon the German attackers is that when the initial points value of their group is *halved* – MINIMUM fire becomes AVERAGE; AVERAGE becomes MAXIMUM; MAXIMUM remains the same. Thus, a force that started out at 100 points suffers increased effects from fire when reduced to half-strength – this reflects (in its effects) a possible lowering of morale through losses.

At the bridges – the attackers fail if a bridge is blown; if on the Belgian side of the Canal they are wiped out, if still on far side, there is little they can do.

The attackers have succeeded when they get across the bridge 28 men (half their full strength) or a reasonable number from a depleted group.

At the fort – the attackers have succeeded when all cupolas containing guns, etc, are blown when it is accepted that they are subsequently penetrated by attackers. This is done by men staying *on* the cupola and placing charges; this takes 1 man *four* moves to do; 2 men take *three* moves; and 3 men take *two* moves – without a break in their occupancy.

Apart from its pocked and battered entrance and a casement still mounting a gun, Eben Emael remains buried under its grass- and tree-covered ridge bordering the Albert Canal, just as in 1940. Considering imaginatively what took place there and viewing the hundred odd graves of the Belgian defenders of Canne Bridge is sobering, but it should not be allowed to depreciate a truly daring and remarkable feat of arms and a landmark in military history.

The Fog of War set up by the dust of a cavalry in the Punjab. The 16th Lancers break through Sikh squares at Aliwal in January 1846 (Figures Willie 30mm)

THE FOG OF WAR

One of the paramount – yet most ignored – elements of warfare in the days of black powder was the almost impenetrable 'fog of war' that enveloped the battlefield from the moment the first volley or bombardments began. Because of it, troops – particularly fast-moving cavalry – were able to loom up out of the fog and strike devastatingly into flank or rear of an unsuspecting unit. By *not* simulating such aspects of 17th, 18th and early 19th century warfare, not only is the wargamer fighting in a quite unrealistic manner, he is also forfeiting one of the most interesting and realistic means of bringing his tabletop battle to life. What follows is a relatively simple method of simulating the 'fog of war', with the added bonus of creating a definitely lifelike appearance to the battlefield.

Across the frontage of each unit of firing soldiers or battery of guns is placed a length of cotton wool, at least equal in height to the figures themselves. The placing of this cotton wool 'smoke' conforms to the following rules:-

On the first volley smoke is placed – 2″ out from gun barrels and covering the exact frontage of the unit or battery of guns.

On a second volley being fired, a further section of cotton wool 'smoke' is placed 4″ out from the gun barrels and extends over the exact frontage of the unit or battery **plus** 2″ more at each end of the frontage.

On a third volley being fired, a section of cotton wool 'smoke' is placed 6″ out from the end of the gun barrels and covers the frontage of the unit or battery **plus** 3″ extra at either end of the frontage.

After that, if volleys continue to be fired, the smoke remains at 6″ but if a game/move takes place without a volley being fired then the smoke placed at 2″ is removed. Subsequently if a second non-firing move follows at once then the 4″ is removed. Of course, on firing being renewed then the 2″ smoke and the 4″ smoke is replaced.

If a system is being used where one gun represents a battery then the battery frontage is considered to be 6″.

When a unit or battery of artillery fires into (through) such smoke then it **must** fire **directly** ahead without aiming. Only that part of the target which comes into the area covered by fire will take casualties. In addition, the effect of firing is reduced by one-third. An added refinement to simulate the attacker's difficulties in penetrating the smoke clouds, and also producing authentic firing prob-

One of the earliest battle photographs, taken by Matthew Brady at the Battle of Antietam in 1862 during the American Civil War. It was titled 'The Battle Fog at Antietam'

lems, is to employ a DEVIATION Device. This is a length of dowelling equal to the longest artillery range, marked in inches, plus a short 3/inches piece of hardboard or card, marked thus – 1 2 3 4 5 6 – painted alternatively red/white.

To use, place one end of the stick against the unit/gun firing or charging, with the stick laying over the numbered card, then throw a normal 1-6 dice. If the dice falls 1 then the stick is moved left so that it lies across the red 1 section of the Deviation-stick; if it falls 2 the stick is moved to the left to lay across the white 2 of the Deviation-stick; if it falls 3 the stick is moved right to lay across the white number 3 section of the Deviation-stick and if it falls 4 the stick is moved further right to lay across the red number 4 section of the Deviation-stick. If the dice falls 5 or 6 then the charging-unit are on target and the stick is left in position laying directly over the 5 and 6 blue section of the Deviation-stick. As soon as the on-coming enemy reach the first smoke in their move then their path must incline to exactly follow the direction now taken by the stick. This may well mean that they completely miss the awaiting enemy thundering past on their right or left flank and may (even) smash into another unit or, on the other hand, they may reach the end of their charge-move and be left wallowing in space to await some unpleasant reaction next move.

When being used to simulate wild firing, the same method is employed and the effects of fire are taken by whatever unit happens to be under the numbered section of the deviation-card – if there are none, then the fire is wasted.

Every line of 'smoke' laying in front of a unit or battery of guns ADDS 2″ to the charging-distance of the enemy coming from the far side of the 'smoke'. Thus a force 9″ away with a move-distance of 8″ (which would mean that, under normal circumstances, they could not reach the unit they wished to attack) can now get to them because of the 2″ bonus added to their 8″ move. When the commander of a charging unit announces that he intends coming through the smoke, this 'freezes' the unit being charged, who may make no move or fire at the attackers, nor may any supporting units make any move resulting from their knowledge of the impending charge.

It can get very wild and woolly, often with some acrimony and argument – it is not a bad idea to have an umpire in games where 'smoke' is being simulated!

4
THE SAND OF
THE DESERT . . .

. . . is sodden red.
Red with the wreck of a square that broke;
The Gatling's jammed and the Colonel dead,
And the Regiment blind with dust and smoke.

From Sir Henry Newbolt's *Vitai Lampada*, written at the end
of the last century, those lines referred to that most evocative
of all Colonial Wars – the Sudan Expedition to get Gordon out
of Khartoum – and the Battle of Abu Klea, fought on 17 January
1885. The 'Colonel' was 'The True Blue' Fred Burnaby of the
Blues, in the Desert on sick-leave, fighting outside the square
with a double-barrelled 12-bore loaded with pig shot; and it
was a Gardner gun that jammed, not a Gatling. A mere four
lines but sufficient to stir the imagination and suggest what it
was like to be a British soldier in a small unsuitably equipped
expedition taking on vastly superior numbers of Dervishes,
Sikhs, Zulus, or a host of other warrior-races, fanatical fighters
in their own completely alien terrain when annihilation was the
only alternative to victory. Incomparably colourful through the
amazing variety of opponents, Regulars and native auxiliaries;
by the varied dress, weapons, terrain, leaders and backgrounds,
Colonial Warfare is tailor-made for wargaming. Providing scope
for dashing – almost outrageous rules – it is the territory of the
audacious wargamer, the chancer or gambler, the player whose
nature and temperament is unsuited for more orthodox modes
of warfare. This wargamer will be in his element controlling large
numbers of lightly-armed, highly mobile native warriors darting
around the wargames table. Not for such a player the complaint
of inequality that discourages wargamers from leading native
armies; rather he skilfully employs every inch of movement
bestowed upon them by the rules, always remembering how the
Afghans won at Maiwand, the Boers at Laings Nek and the Zulus
at Isandhlwana – feats well within reach on the wargames table.
Or he might fancy himself in the heroic role as commander of a
far smaller force, effectively massing his extra firepower while
demonstrating the coolness, courage and control of a Roberts
or a Wolseley.

Throughout the Victorian years, men from every County in
Britain, in redcoats or dusky khaki, marched in slow-moving
columns, elephants jostling camels, bullocks plodding with
donkey and yaks, mules carrying the little mountain-guns

A wargame attack on a North West Frontier tribe's village

*Victorian Colonial Campaigns –
the first Sikh War 1845/6*

The Boer War 1899–1902

Like the Spanish-American War, the Boer War cannot be considered as a typically Colonial campaign because white men were facing each other with relatively modern weapons. Although bloody, the war proved little of military value except to prove that modern rifles in the hands of skilled marksmen had ended the day of daylight tactics involving formed bodies of soldiers in the open. On the other hand, the expansive terrain, the climate and the absence of modern transport perpetuated a number of false conclusions as to the role of cavalry in modern warfare.

Although untrained in the military sense the Boers were formidable opponents. Exceptionally mobile on their horses, the South African farmers fought mainly on foot, usually from concealed defensive positions. They formed themselves into commandos, small hard riding groups of bearded, slouch hatted farmers with cartridge belts around their shoulders. These commandos would swoop down upon British camps, shoot them up and be away before any resistance could be organised against them.

It was a war of movement in which formal drill was little use against crackshots hidden in invisible trenches. This meant that the British had to adopt Boer methods and mounted infantry units were hastily formed.

The later stages of the Boer War consisted of guerilla warfare which was slowly defeated by a systems of garrisoned block
continued

clattered and jingled. Under dashing commanders such as Lord Charles Beresford, boisterous, straw-hatted sailors and Marines dragged Gatling and Gardner guns through the sands of the Sudan; during the Indian Mutiny, Captain Peel of HMS *Shannon* and his sailors manhandled huge naval guns to Cawnpore and Lucknow, and the 4.7 naval guns brought ashore during the Boer War and the Boxer Rising provided a piquant note to an already colourful period. In this heyday of Colonial Wars, the British soldier took on – and usually defeated – Afghans, Africans, Afridis and Asians; Baluchis and Boers; Fanatics and Fuzzy Wuzzies; Kaffirs; Kings, Princes, Rajahs and Chiefs; Mahrattas, Maoris, Mandarins, Monks and Mullahs; Pathans, Sikhs, Zulus and even rebellious Canadian half-breeds.

Some of these expeditions were by disciplined soldiers against savage or semi-civilised races, aimed at adding their territory to that of the Crown, campaigns of annexation usually against the warriors of a specific ruler as in Sind in 1843, against the Sikhs in 1846 and 1848, or to curb King Thebaw of Burma in 1885. After hard fighting, operations frequently developed into operations of suppression involving ambushes and guerilla warfare, resulting in stern and savage reprisals. Sometimes an expedition went out to avenge a wrong or wipe-out an insult – as in the Abyssinian Expedition of 1868; or to overthrow a dangerous or troublesome enemy; although not the original intention, frequently they ended by bringing vast new territories under British rule. Most of the expeditions on the Indian Frontier were in this category, many resulting in the offending territory being annexed; it was also the case in the Zulu War of 1879. On occasions these campaigns turned into affairs of expediency fought for political reasons, like the two Afghan Wars and the Egyptian War of 1882. In a sense, Colonial Forces operated like Fire Brigades living

54

within hearing of an alarm-bell whose strident clangour sent them anywhere to quell an outbreak.

Colonial small wars were campaigns against Nature as much as the enemy; in climates quite unsuitable for trained European soldiers, losses from sickness were often greater than by fire and sword. Bush, desert and jungle imposed unusual restrictions on commanders; movements and supplies were greatly handicapped by lack of communication, and the armaments and equipment necessary for tactical superiority burdened them with non-combatant services, bases, and lines of communication. The strength of civilised forces contained a built-in weakness as – backed by all the resources of science, wealth, manpower and navies – organised regular expeditions found themselves at undoubted tactical and strategical disadvantage in these small wars. Primitive conditions, the singular features of the theatre of operations, relatively little knowledge of the fighting qualities of the enemy and his methods of warfare and weapons, all combined to make Colonial warfare quite different to the regular warfare of the period, where both sides were governed by accepted common rules. All manner of opponents were met, all fighting in a different way – some being trained and organised like Regular troops as the Egyptian Army of Arabi Pasha in 1882; and the Sikhs of the Punjab, arguably the best organised and hardest fighting native army encountered during the Victorian period, trained by experienced European

houses at regular intervals, dividing the country into relatively easily controlled sections. The basic situations of the Boer War can provide a most stimulating tabletop campaign in which one force, wholly mounted, can combat against a regular army of infantry, cavalry and artillery, the increased mobility of the mounted men enabling them to complete vast map-moves before picking their battle ground.

Cavalry against Zulus – Battle of Ulundi (Zulu War) 1879

mercenaries, their artillery ranked among the best in the world at the time. The Zulus were a well controlled and organised army, capable of carrying out manoeuvres with order and precision, although using primitive weapons; and in the Indian Mutiny of 1857, the enemy consisted of vastly superior numbers of soldiers trained and disciplined in the British manner, acting cohesively as formed regiments, but badly led.

All fought quite differently to the no-less formidable Afghans, Ashantis, Boers, Chinese, Dervishes and Maoris, while Sudanese and the Afghan Ghazi lacked the discipline of the Zulu but fought with a bravery and recklessness rendering useless tactics successful against conventional opponents and forcing a British return to the long discarded Square formation. The Kaffirs of Africa possessed lower levels of courage and were poorly armed, but proved most difficult to subdue when utilising natural advantages of bush and jungle. The turbulent North West Frontier of India, a running-sore for a century, provided experience of active service conditions to generations of British soldiers, otherwise dragging out their lives in Britain, around grimy barracks or ancient forts and castles. In their familiar harsh terrain, belligerent, brave and tactically brilliant tribesmen, crack-shots with stolen or 'home-made' rifles, flitted like fanatical ghosts.

During the late 1870s and early 1880s, over a three years' period and in one single quarter of the African Continent, British troops came successively into conflict with the astonishingly different methods of warfare employed by Kaffirs, Zulus and Boers.

A Colonial wargame involving North West Frontier tribesmen

At exactly the same time the regular army was heavily committed on the North-West Frontier and in Afghanistan. Unless the Regulars adapted their fighting-methods to cope with this variety of foes and their different modes of combat, unforseen difficulties and even disaster resulted. The overwhelming Zulu victory at Isandhlwana in 1879 was directly due to a total misconception of the enemy's tactics (made worse by lack of availability of ammunition); reverses in both Boer Wars arose from lack of cavalry, the one essential arm in that type of warfare.

Coping with this is part of the fascination of wargaming in the Colonial period, necessitating a knowledge of the varying fighting-styles and adapting game-conditions and rules to reproduce them. It would be quite out of historical context for Boers to charge en masse in Zulu style, nor would Zulus flit around in bush and jungle like Kaffirs and neither of the coloured groups were armed with rifles that allowed long-range marksmanship in the Boer manner. From the British point of view, the necessary massing in close formation to throw back onrushing Zulus by sheer weight of firepower would be fatal against the accurate shooting of the Boers, when dispersal, cover and perhaps fire-and-support methods would be essential.

Of the sixty or so wars and campaigns fought by the Victorian soldier, only one was lost – that against the Boers in 1881 – through misapplied tactics that no self-respecting wargamer would perpetrate on a tabletop battlefield. All British commanders of the latter part of the 19th century were not of the standard of the ubiquitous Generals Roberts and Wolseley; there were plenty as hidebound as General Colley at Laings Nek in '81.

When wargaming these wars it has to be over-emphasis on the power and effect of modern weapons that gives the small British forces any chance against vastly superior numbers of lesser-armed natives. In such moments of stress as combat, whether or not a unit or army stands firm is determined by its *morale*; this is as important in wargaming as in real life so the quality has to be grafted upon inanimate model soldiers. A prime means of bestowing upon smaller, disciplined forces the ability to defeat much larger groups is by giving them a higher morale rating and allowing that of tribesmen to fluctuate so that they can display extreme bravery tinged with the possibility of breaking and running, and being impetuous so that they courageously attack at tactically unsound times – simulated in wargaming by 'uncontrolled charges'. More than European troops, native tribesmen often find their morale affected by the death of a chieftain or leader. When there is a known parity of armament (the Sikhs in 1845–6 and 1848) or when fighting European opponents such as the Russians in the Crimea and the Boers in 1881 and 1898–1902 – then only morale differences can make wargaming viable. Lack of numerical balance between opposing forces becomes practical through allowing the smaller, disciplined force advan-

Scales

The major difficulty when wargaming in the modern era is that artillery ranges are measured in miles rather than yards! Even World War II tank battles commenced at ranges of 1,500yd (1,370m) and, in the clear air of the desert, this increased to over 3,000yd (2,700m). Taking the lower figure of 1,500yd (1,370m) at the common scale of 1:72, the range is approximately 60ft (18m), which cannot be scaled down to the tabletop battlefield. Even at the very small scale of 1:300, ranges of 1,000–3,000yd (900–2,700m) have to be represented by 10–30ft (3–9m).

For 1:72 scale wargames the length of the table should be equal to 2,000yd (1,800m), the maximum range at which tank battles normally took place. With the average 6ft (2m) long table, a suitable ground scale will be 1:1,000. If each game-move reresents 30 seconds of real time then an infantryman will move 2in (50cm) per game-move; if it seems slow, remember that most modern infantry rode in trucks, half-tracks or on tanks. At this scale, time-relationship 1 mile per hour is equal to ½in (12mm) per move (2kph=17mm per move).

Colonial wargame around a native village

Making them Look Real – Painting Model Tanks
In World War I no fixed colour pattern appears to have been adopted by any of the combatants. The first British tanks were painted battleship grey, becoming progressively greener as the war progressed. Towards the end of the war British tanks appeared in a camouflage pattern, which included light blue among the dark browns and greys. French tanks were painted in a camouflage pattern of grey, yellow, green and brown, each colour delineated by a thick black line; and some were painted in a green and brown pattern, similar to British aircraft in World War II. German tanks were in a mixture of green, lilac
continued

tageous rules to simulate their superior tactical formations and enhanced firepower. The natural mobility of natives is simulated by longer move-distances; their ferocity and fanaticism at close-quarters reflected by giving superior values in hand-to-hand combat (meleés), once they have surmounted the volume of fire they must encounter on their way in. Native foes should be granted facilities within wargame rules and conditions for surprise and ambush in their own familiar terrain.

Natives hated cavalry, particularly Lancers (who were also dreaded by the Boers after their early experience against them at Elandslaagte in 1899) – rules should make it difficult for natives to stand against cavalry and, if they break and are pursued, to have a much reduced defensive capacity. Tribesmen disliked artillery fire, so rules should reflect this by forcing them to break easily when under fire, or to back-off out of range or seek cover. Colonial conditions and terrain usually allowed light artillery to be used and Britain won her Empire with the 13pdr 8½cwt, 3in calibre field-gun and the mountain guns, carried in separate parts on mules. The hillmen of India's North-West Frontier said that what they feared was '. . . not the child-rifle but the devil guns which killed half a dozen men with one shell which burst and threw up splinters as deadly as the shots themselves.' These mountain batteries repeatedly feature in Colonial small wars and their officers claimed that they could go anywhere that a man could go! In 1880 the famous screw-gun (immortalised by Kipling) was introduced with the barrel made in two pieces, which screwed together.

Wargaming does not consist only of moving miniature armies on tabletop terrains, it includes the collecting and colouring of regiments of infantry, squadrons of cavalry, batteries of guns, and hordes of tribesmen, which takes time and tends to discourage the belligerent wargamer who wants to get on with the battle. Perhaps the Colonial wargamer has the best of it – painting smaller numbers of Regular troops either in rapidly-applied khaki or a monotony-defeating variety of uniforms, while native robes (or lack of them) can be handled with complete lack of uniformity in a 'please yourself' attitude. The wise wargamer forms his forces of the greatest variety of types, and the Colonial adherent has a great number of native auxiliary units at his disposal, who are more interesting to paint. Most Colonial campaigns were fought largely with native troops raised from tribes and groups in annexed territories – in their Empire-building days, the British were second to none in their ability to organise and successfully maintain forces recruited from people recently conquered – there are few recorded cases of them failing to co-operate harmoniously and loyally in the years that followed. Initially these Askaris, Hausas and Sepoys were used in their own areas, later in other parts of the Empire – Indian units fought in the Sudan, in Uganda, Nyasaland and Somaliland; West Indians went to West Africa; Central African units to Ashanti, and Nigerians and native troops from the Gold Coast fought in Tanganyika; and Kitchener's Dongola Campaign of 1896–8 was carried out with two British brigades alongside four Egyptian and Sudanese brigades. The latter were recruited prisoners from the Mahdis armies who, after being trained and bullied by sergeants of their own race and colour, set about fighting their former comrades with great enthusiasm! Native units should not be underestimated so far as fighting ability and morale are concerned – just like European units some were better than others, their quality depending upon equipment, standard of training, experience and European commanders – these factors should be taken into account when formulating rules and conditions, just as for European units. Native units were usually issued with obsolete equipment, being armed with Snider rifles when British troops had Martini-Henrys, passed on when the Lee-Metford arrived, which in turn was handed over to native units when the Lee-Enfield came on the scene.

The exciting range and scope of Colonial Wars can be extended to wars fought by Foreign Powers in other parts of the world, or in periods outside the Victorian years – British and Indian troops were fighting native armies in the 17th century, and the French-and-Indian Wars of the Seven Years' War are an unusual and colourful project. In a wider sense, Napoleon's 1798 invasion of Egypt was a Colonial War and the wargamer will be attracted by French squares battling against hordes of gaily coloured and wildly courageous Mamelukes, with flotillas

or purple, and pink and brown smudges.

Between the wars the British employed a medium-green livery, which was rarely camouflaged; armoured cars operating in desert areas were sand colour. French tanks of this period were camouflaged in a green and two-shaded brown smudged mixture, sand colouring being added when appropriate to the terrain in which they were operating. The few German tanks of the inter-war period were painted in varying shades of grey; other countries tended to use a mixture of grey and green as a camouflage pattern.

At the outbreak of World War II these colours prevailed, and no new patterns emerged until the Desert War in North Africa, when British and American-made tanks bore a basic sand colour with very little camouflage. The British also used a pinkish sand colour with battleship grey and brown stripes running in various directions. The Long Range Desert Group vehicles were painted sand colour with darker tan markings. Italian tanks and armoured vehicles were a basic sand colour with scattered brown smudges. German vehicles in the Desert were sand-coloured.

World War II Russian tanks were painted either grey or a uniform light olive green, presumably dependent on the time of year, with the grey colour predominating during the snowy conditions of the Russian winter. German winter camouflage was a mixture of light greys and greens; otherwise it was dark tan and brown. The Germans employed Zimmerit, a roughened surface paint (giving the appearance of combed-concrete), to prevent Russian sticky-mines adhering to the tank armour. During the Allied invasions of Italy and
continued

France the basic colour of all vehicles was dark olive green, modified to suit terrain by adding tan. As the Eastern Front moved away from the snowy plains of Russia, tank colours closely resembled those on the Western Front. The Japanese and British tanks in Burma were patterned in striking combinations of dark shades of yellow and light shades of green.

Since World War II the basic colours of armoured forces have been variations in green and green and grey. The ease of spotting armoured vehicles by sophisticated equipment has made camouflage less valuable, but the firing of guns still depends on the eye of the gunlayer, so camouflage has its value. The Russian and Iron Curtain countries tend to use a mixture of dark grey and olive green for their tanks; and NATO forces use two shades of green in a pattern somewhat similar to that of World War II aircraft. The vehicles of the Arab-Israeli conflicts were usually sand colour. Modern armoured vehicles are sometimes covered with washable distemper paint, so that it only takes minutes to recamouflage a vehicle to suit changed ground conditions, as the paint can be washed off with water and a solvent.

There are a number of extensive ranges of colours to match the exact shades of World War II model armoured vehicles; also 'weathering' paints to simulate dirt, grease and mud.

of French and native boats fighting nearby on the River Nile. Colonial campaigns were fought during the 19th and early 20th centuries by France, Germany, Turkey, Russia, Spain, Portugal, Belgium, Italy, Japan and the United States, in such far-off places as China, Persia, Cuba and the Philippines. The French fought colourful campaigns in North Africa from about 1836 onwards, then in Tonkin and Indo-China, and parts of West Africa, employing many exotic military formations – Foreign Legion. Chasseurs d'Afrique, Spahis, Zouaves, Turcos, Marines and the Colonial light infantry, camel and mule mounted troops, and native Algerians, Senegalese, Malagasys, Hausas and Indo-Chinese. Coming late into Empire-building, Germany fought some fierce campaigns in East Africa, Togoland and the Cameroons, using a colourful variety of troops including Marines and European Schutztruppen, mounted infantry, camel corps and different types of Askaris. Their opponents, the Hereros Ovambos, Masais and East African natives are noticeable by their absence from the figure makers' catalogues, but the resourceful wargamer will readily adapt and convert other available native warriors to suit.

Although the passing of Queen Victoria saw the beginnings of a close-season for Empire-building, the British Army was periodically engaged in punitive operations on the North-West Frontier of India as late as during World War II. An interesting and little-used wargaming period can be the conflict with fanatical followers of the Fakir of Ipi, a Holy Man who flourished in the 1930s, when County Regiments, supported by armoured cars and light tanks, went out on many expeditions in those same barren hills fought over by their fathers and grandfathers. There were a few airplanes involved, ancient bi-planes whose reconnaissance and bombing revolutioned the style of warfare – it was generally considered by the soldiers that they spoiled the sport!

Living as we do in an age when children's golliwogs have fallen into disfavour through their implied racialism, it is quite likely that accusing fingers will be pointed at tabletop representations of historical occasions when natives defending their homelands were massacred in large numbers. Perhaps there is some mitigation in the knowledge shared by most wargamers that, in spite of suffering at least a hundred casualties to every ten European, these ill-armed natives invariably revealed incredible courage and dignity, being defeated solely through being overtaken by the March of Civilised Progress. Wargames rules thoughtfully reflect their courage and, as there is no enjoyment in playing a game where the same side invariably wins, the best man always triumphs in these tabletop affairs – and it is the native tribesman more often than not!

DOWN AT THE WARGAMES CLUB No. 1

The Gang who formed the Wargames Club knew each other pretty well, being fully aware of each other's strengths and weaknesses they are always ready to wind-up anyone for the sake of a laugh. When things are quiet it is'nt difficult to get an argument going on the respective merits of figure of board wargaming, on scale, fantasy, uniforms, etc., and, as in most groups, there are one or two well-known 'stirrers', like Pete and Alan. The other night Pete came into the Clubroom and showed us some photographs he had taken on his recent motoring holiday in Spain. He passed one to Billy Phillips: 'Know where it is, Billy?' Everyone watched closely wondering what was going to happen next, because it was well known that Billy firmly believed British Military history began with and revolved around the 95th Rifles and the Light Division in the Peninsular, and that he could always be got going by anything concerning them. Quick as a flash, Billy said: 'It's the Lesser and Greater Arapiles on the Field of Salmanca . . . fought on 22 of July 1812.' Then Pete started raving about how marvellous Packenham's 3rd Division had been, and how the 4th and 5th Divisions came over the ridge and helped roll up the French – '. . . 40,000 Frenchmen beaten in 40 minutes!' 'But, Billy . . .I've often wondered where the Light Division were whilst all this was going on?' Billy choked and went red: 'You know perfectly well where they were . . . they were on Wellington's left flank along with the 1st Division . . . the best disciplined and controlled divisions the Duke had . . . waiting to take up the pursuit when the French began retreating!' Pete looked thoughtful: 'Is that so? Funny, I don't remember hearing much about the French getting badly beaten-up in the pursuit!' Billy waved his arms like he always does when he's excited: 'That was because a Spanish force under D'Espana placed by Wellington in the castle at the bridge at Alba deserted their posts and allowed the French to get away!' Pete laughed sarcastically and the argument got fierce and furious, until the President came over and said they were disturbing the rest of the wargamers.

To quieten things down and keep the peace, Toby hastily told us about a film he'd seen on a friend's video the other night: 'It was supposed to be the British Army on their way to the Battle of the Alma in the Crimea, and they had this lot marching with a high knee action and arms going across the chest . . . like foreign soldiers . . . it was a damned insult to the British Army!' 'That was the film *Charge of the Light Brigade*, wasn't it?' said Billy, forgetting about the 95th for the moment: 'They used the Turkish Army as extras.' Then Fred said someone he knew who'd worked on the film told him that the British Director, offended by Turkish cavalrymen playing the Light Brigade slumping in their saddles, lashed lengths of broomstick up their backs to keep 'em upright!' Billy laughed bitterly: 'That was the same director who insisted on one of the British cavalry regiments in the charge wearing scarlet overalls and when the military adviser protested that they wore blue, he said it was HIS army and he could dress them how he liked!'

'In the film *Waterloo* all the Allied and French units were formed of real-life soldiers of the Russian Army . . . there was one scene where the so-called Gordon Highlanders had their tartan stockings down around their ankles . . . they did'nt look anything like Scotsmen!' Fred expressed sympathy and said: 'Of course the thing that spoils all military films is that there are never enough soldiers in the regiments . . . they can't afford to pay sufficient extras to make a unit look realistic. They show you a spaced-out column of about 40 men marching in three's and it's supposed to be a full battalion.' A guy we hadn't seen in the Club before sniffed and broke our rule of never bringing politics into a meeting: 'Obviously they can't use as many men as took part in the real historical battle . . . the only time enough money can be found for that is when one country makes up its mind to kill all the soldiers of another country!' Not fancying the turn of conversation, Fred said: 'I thought the Russians had pretty large units in their version of *War and Peace* . . . made without thought of expense and used as propaganda, I suppose.' Pete agreed: 'I've always thought about the most real-istic film of horse and musket fighting I've ever seen was in *Barry Lyndon* . . . the attack in line was just about how it must have been in those days.'

Don't usually hear much from Chris Blake, known as a deep-thinking lad with a degree in some electronic subject, but he came into the con-versation: 'The nearest you can get to realism is in tabletop wargaming, where a unit can be as big as you like if you can afford to buy the figures.' Fred interrupted: 'And if you can find an opponent with a similar sized army.' Chris nodded: 'Table size means regiments always have to be drastically scaled-down, but that's not the main trouble . . . the real fault of wargaming lies in the fact that the figures don't move of their own volition . . . they have to be moved.' Someone said what we were all thinking: 'Of course they can't move on their own . . . they're only metal or plastic.' Chris shook his head: 'I'm convinced there is a way to make them move on their own!' That aroused a general free-for-all, with everyone talking at once, then someone reminded us of a recent film at the Odeon about a couple of American teenagers who fed vital statistics into a computer, to make their very own glamour-girl. 'Miss Frankenstein, I suppose,' said Billy. Chris smiled: 'I'm working on an idea . . . I'll let you know when I come up with something.'

It made us a bit uneasy, you never know what these electronics wizards are doing – and there were all sorts of queer rumours about the factory on the Industrial Estate. Anyway, the weeks passed and we forgot it until Chris invited some of us round to his lodgings, to see a video film he'd made. It really made us sit up and take notice, this short and not particularly clear clip showing what were undoubtedly blocks of model wargames soldiers of the Napoleonic period jerkily moving towards each other, until blotted out by clouds of smoke as they fired volleys! We saw it through half a dozen times without being really convinced before Billy said: 'They are mounted on a large base with a flange underneath moving in a channel . . . controlled by radio like model cars or aircraft.' Chris was indignant: 'How do you account for the firing then?' 'Some sort of a firework set off by radio signal' said Billy. Chris shook his head sadly: 'You don't know what you're talking about . . . I'm afraid you are out of your element.' 'Alright then tell us how it's done!' 'I'd like to,' said Chris 'but I can't . . . there's more to it than you think, it's part of a much larger affair . . . Ministry of Defence involved . . . Official Secrets Act . . . you know how it is.'

We told the bloke who owned the Model Shop about it and he laughed, saying he hoped Chris would let him have exclusive sale of his patent figures – 'Could be trick photography . . . I once saw a TV film being made of a cavalry charge on a gun . . . by moving the figures and the gunners a fraction each shot they made more than a hundred and fifty frames that showed for a few seconds on the screen and looked very realistic.'

A week or so later Chris Blake came into the Club; he had his arm in a sling. 'Had an accident?' He smiled and gingerly took his hand from the sling and carefully removed the dressings, holding out his hand for our inspection. Both sides of it – back and palm – were peppered with angry little wounds! 'Twenty-five each side . . . count 'em if you like . . . I was playing a solo-game with my army and put my hand in the way of a volley – not a man missed!' We still don't know what to think – there's no way a guy would give himself fifty regularly spaced painful wounds – just to prove a point, is there? Or is there?

5
BY AIR TO BATTLE

Since Man first raised weapons against his fellows there must have been numerous occasions when military commanders – even the Alexanders and the Napoleons – fervently wished that their armies could rise from the ground, fly over the enemy and engage them from the open (or vertical) flank. But men only flew in legends or in the dreams of visionaries like Leonardo da Vinci, so commanders remained restricted to two-dimensional warfare and the search for a weak right or left flank. It remained so until the development of aircraft that, by allowing troops to be landed from the air, endowed them with a highly specialised degree of mobility to open up a third (or rear) flank. The Russians pioneered this new style of warfare, influencing the rearming Germans, who realised that assault-troops using parachutes could be delivered by existing aircraft without need for costly single-purpose 'planes. They reasoned that an airborne force possessed the same mobility and radius of action as the aircraft itself, although delivering troops by air was a one-way system, soldiers arriving on the ground scattered and armed only with light equipment, supplies and ammunition limited to what they could carry from one supply-drop to the next. The Fallschirmjager song: 'Comrades, There is No Going Back!' was a wry admission of limitations of tactical mobility.

Accustomed to playing God over tabletop battlefields, wargamers should be attracted by the chance to do it in another dimension; well within reach of all, it is among the cheapest and easiest forms of wargaming. An ideal club or group project, try this stimulating reconstruction of the two best-known and most evocative of all airborne operations, both worthy of simulation because – in their own way – each was a milestone of this form of warfare. The German invasion of Crete in May 1941 represented the zenith of German Fallschirmjager activities while spelling their doom, and in September 1944 British/Polish airborne operations at Arnhem (Operation Market-Garden) coupled with those of the Americans at Eindhoven and Nijmegen were outstanding examples of Allied airborne co-operation. Formed of numerous smaller actions, both battles demonstrate the fighting-style and tactics employed by para and glidertroops and amply justify the effort of assembling miniature airborne armies, plus the necessary pre-planning and terrain construction.

Space does not allow for sufficient details of either battle, so they will have to be read-up in some good source (such as

A wargaming figure of a British World War II Airborne soldier

63

A wargaming reconstruction of Frost's Airborne Division's defence of the bridge at Arnhem September 1944

Maurice Tugwell's definitive work *Airborne to Battle: A History of Airborne Warfare 1918–1971* (William Kimber 1971) obtainable from any self-respecting Public Library). When reading, consider each phase and seek possible alternatives to the historical trend of events; decide whether your reconstruction is to be a reasonably accurate simulation or a wargaming re-fight. There is so much information available on formations, dispositions, tactics, objectives, timings etc, that it would seem easier to accept known facts and faithfully reconstruct these historical operations, allowing Chance-Cards (more about them later) or dice-throws to supply variations and uncertainty at appropriate points.

The obvious way to reconstruct a battle is for troops to perform precisely the same actions as they did on the day, take the same percentage losses, with the same success or failure. However, that will be an historical exercise and not a wargame, serving only as a demonstration of past events and ignoring hindsight. It is important to consider what *might* have happened in conjunction with what *did* happen; better to follow – in context and chronological order – the original happenings, allowing some leeway but without overstretching the imagination; although you are playing a game with the objective of logically reversing the original result, do not take too many liberties or it will become a wargame played merely for its own sake. To bear a noticeable similarity with real life events, the tabletop terrain must closely resemble the actual battlefield, with the same dropping and landing-zones. But how-

64

ever accurate the miniature topographical features might be, it is pointless to allow the combatants to mill around in a manner bearing no relation to the original conflict – so the scaled-down numbers equivalent to those on the day must be controlled by adequate rules and conditions.

If a club or group activity, call an introductory or briefing meeting to consider the practical aspects of setting up a relatively extensive wargaming operation, and to allocate roles; re-fighting either of these memorable battles can involve many members who will represent the various commanders on the spot, either under the direction of an experienced wargamer as overall commander or, as occurred historically through faulty communications, will be forced to make their own decisions during the course of the battle. The series of simultaneous small actions that made up Crete and Arnhem are best wargamed by fighting each one on a separate wargames table, ideally in the respective homes of the participants. Thus for Crete, Wargamer Charlie will set-up the area around Maleme Airfield; Tom the Retimo area; Gavin is responsible for the Heraklion area. Unlikely to be concluded in a single evening, each game will continue until conclusions are reached; with progress kept secret from participants in the other games. Ways and means of explaining to wives and/or parents this invasion of their privacy must be considered a personal matter and outside the scope of this article! If this system is not possible decide the order of fighting each separate area without worrying about chronological order, considering the whole project on a 'league-table' basis, with points allocated for winning or drawing a battle, totalled at the end to decide the overall winner.

Ideally, a Co-ordinator will mark-up the course of operations on a large map of the entire area, copied in larger scale from maps given here, or commercial maps; and each wargamer will have a large-scale map of his personal area of operations, kept up-to-date with troop dispositions and movements. These 'sector' maps are drawn wargames-style, with topographical features condensed to fit table-space and with irrelevant features omitted. Maps are marked by covering them

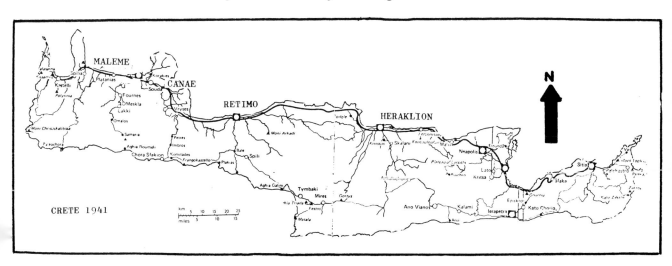

CRETE 1941

with transparent plastic-sheet and using coloured wax pencils. Occasional meetings to discuss progress (but without revealing current successes or defeats) will stimulate interest.

Crete and Arnhem, fought as wargames, are 'sectional battles'; in the former German East, West and Centre groups fought quite separately; Operation Market-Garden saw the 82nd and 101st US Airborne Divisions fighting south of Arnhem simultaneously with Frost's defence of the bridge and attempts from northern landing and dropping-zones to join him. The Co-ordinator must keep a Time Chart on which is co-ordinated manoeuvres and stages with wargames-moves, as they occur; only in this way is it possible to keep check on such simultaneous and pro-grammed situations taking place on maps, with different forces moving along various routes, or attempting movements to bring troops onto the tabletop battlefield at intermediate stages of the conflict. Both battles were radically affected by the clock, the anticipated time of arrival of reinforcements and supplies being of vital tactical importance, Commanders, perhaps unaware of the exact location of detached parts of their force, will need to communicate by messenger or radio; such facts must be recorded on the Time Chart so that the exact time of the arrival of the message is known and the units affected cannot react until the message arrives. Non-arrival or delay can realistically alter the course of the reconstructed battle.

Chance-Cards are originated by the Co-ordinator, giving him a chance to blow off steam or display his low-cunning; they consist of a set of cards marked with varying conditions and alternatives, peculiar to each situation, that *must* be followed by the recipient of the card. Their rulings introduce factors materially affecting the battle, even its result, by posing tactical, physical or psy-chological factors to be tackled by the commander drawing the card. For example, at Arnhem the glider carrying Gough's jeeps (planned to dash for and hold Arnhem bridge) was destroyed – had this *not* occurred, then the whole outcome of the battle could have taken a different course. A Chance-Card would set-tle this one way or the other, but they must be used sparingly and with discretion or the historical aspects of the battle are unnaturally distorted. Should the Co-ordinator be disinclined to undertake such a task, the situations can be debated on the spot, as they occur, and settled by the throw of a dice – by far the simplest method of simulating the fluctuations of fortune in war.

Para and glider-borne troops, with their carrier, tug and transport aircraft, transform conventional wargaming into fast-moving affairs where lightly-armed elite troops – with an open-ing advantage of surprise – combat conventionally-armed forces backed by armour, artillery and sometimes aircraft. Such games can begin with the airborne forces already on the ground, their arrival an acknowledged fact; or the actual para-drop or glider-landing can be imaginatively simulated. The means of doing this can be either by simply placing groups of paratroopers on the

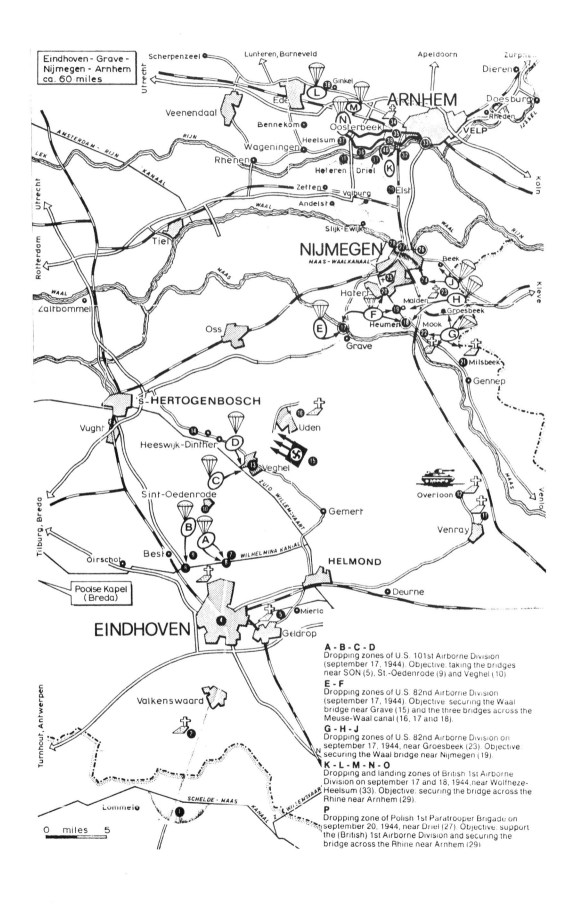

Eindhoven - Grave -
Nijmegen - Arnhem
ca. 60 miles

A - B - C - D
Dropping zones of U.S. 101st Airborne Division
(september 17, 1944). Objective: taking the bridges
near SON (5), St.-Oedenrode (9) and Veghel (10)

E - F
Dropping zones of U.S. 82nd Airborne Division
(september 17, 1944). Objective: securing the Waal
bridge near Grave (15) and the three bridges across the
Meuse-Waal canal (16, 17 and 18).

G - H - J
Dropping zones of U.S. 82nd Airborne Division on
september 17, 1944, near Groesbeek (23). Objective:
securing the Waal bridge near Nijmegen (19).

K - L - M - N - O
Dropping and landing zones of British 1st Airborne
Division on september 17 and 18, 1944,near Wolfheze-
Heelsum (33). Objective: securing the bridge across the
Rhine near Arnhem (29).

P
Dropping zone of Polish 1st Paratrooper Brigade on
september 20, 1944, near Driel (27). Objective: support
the (British) 1st Airborne Division and securing the
bridge across the Rhine near Arnhem (29)

How to Play an Ambush on the Wargames Table

A feature of Colonial warfare was the ambush, usually laid by natives but military history shows many examples set up by disciplined troops. Because of the God-like position of the wargamer towering over the table, nothing hidden from his piercing gaze, a satisfactory ambush is a most difficult thing to set up on a tabletop terrain. One method is to mark a counter or small square of brown or green card with the actual number of the men in the ambush party – there should be a counter for each party so that men can be placed on either side of the road or in different positions. An advancing force can send scouts ahead of them, when the actual figures representing the scouts come within 3in (7.5cm) (or closer if the ambush is round a corner for example) AND the wargamer himself can actually detect the hidden counter, then the wargamer setting up the ambush gives him a piece of paper on which is marked the number of men in the ambush, plus or minus (as he wishes) 25 percent of their actual strength. The scout then attempts to return to his main body but if he is killed or fails to return then the paper is destroyed before it has been read by the wargamer to whom it is handed. The troops must continue onto the ambush area where the ambushers have the opportunity of firing first at them.

If the scout manages to get back to his main body, then they are alerted and their commander can now unfold the paper giving *continued*

German Airborne forces with a light gun in action in Crete in May 1941 (figures, AIRFIX AND ATLANTIC)

wargames table at predetermined landing-points, or by causing to flutter to the table-surface flimsy paper shapes, marked to represent variously-armed troops. This, and other methods, are described at length in other appropriate books by the author; glider-landings are described elsewhere in this book (see Chapter 3).

If the battle is to bear a resemblance in more than name, the tabletop armies must be an accurate proportion of the real life forces – a realistic scaling-down of real life numbers and types is essential. In small operations the wargamer can use a 'one-for-one' scale – the same numbers of model soldiers on the wargames table as took part in the original action, but the large numbers involved at both Crete and Arnhem make this a bit ambitious. Perhaps the most satisfactory solution is 1:20 with one figure on the table representing 20 in real life, so that the 4,300 German paratroops who fought at Crete for the first eight hours will be represented by 215 figures. Or, ignore actual numbers of men and concentrate on units and formations – instead of one man on the wargames table representing 20 men in real life, allow a tabletop unit (even if only 20 men) for each formation historically present. This is facilitated by the battles being made-up of a number of smaller actions, each requiring a separate wargame – which allows *all* available model soldiers to be used for *each* game. This is much better than saying . . . 'I have got 500 British and American paratroopers to be divided into the various units of the British 1st Airborne Division at Arnhem, the 101st US Airborne Division at Eindhoven and the 82nd US Airborne Division at Nijmegen.' Such a method broken-down into battalions means that each will consist of about two figures! If *all* 500 are used to represent single phases of the battle in turn, then formations of more realistic sizes can be fielded.

And where do all these little tabletop warriors come from? Well, that brings us back to the earlier claim that this is an inexpensive branch of wargaming with readily available model

soldiers of all types fighting these operations. Undoubtedly the cheapest method is to buy 1:300 scale metal micro-figures and accessories (Heroics & Ros Figures; Unit 12, Semington Turnpike, Semington, Trowbridge, Wilts BA14 6LB) currently on sale at £1.15p per pack of 50 figures; from the same source can be obtained the right models of gliders and towing aircraft for both Crete and Arnhem. Of course, at a scale of 1mm = 1 foot or 3mm = 1 metre, they might be considered a bit on the small side; although that allows an accurate battle to be fought on a compact terrain with forces approaching in numbers those of real life. And they have to be painted, but that is not too much of a chore as they can be given an overall coat of khaki, grey or whatever uniform is being represented, faces and hands flesh coloured and weapons black.

Obtainable from major hobby-shops are 1:72nd HO scale figures in plastic of all necessary types of airborne troops, German, British and American, made by either ESCI or Airfix; currently they sell for under £2 for a box of fifty figures, well moulded in a useful variety of positions. As with the smaller figures, painting is not really a problem, perhaps less so with these plastic models moulded in plastic which is the basic colour of their uniform, so that only faces, hands and weapons need painting.

Primed with information about the battles, armies purchased, painted and poised ready for action – how do their wargamer-commanders, agog with militant enthusiasm, go about manoeuvring them authentically on tabletop terrains resembling Crete or Arnhem? The whole business is controlled and governed by rules and conditions formulated to produce actions and reactions conforming to those employed in real-life at the time. Obviously, to cover every minute detail complex sets of rules will be required, needing to be constantly consulted and probably involving lots of paper work – all of that is the death-knell of enjoyable wargaming. So, try the other extreme and compile rules that are basic and simple yet – with goodwill on the part of all players – give a fast-moving, reasonably realistic game.

Move-distances and weapon-ranges must be formulated to suit the size of the wargames table, remembering that too-short move-distances slow the game up. To simulate the superiority of being selected men, highly-trained and battle-hardened, airborne troops should be given fractionally longer move-distances; bestow upon them the privilege of firing first in confrontations, to simulate their superior speed of reaction. The diversity of the 20th-century infantryman's arsenal is represented by allowing *all* troops the faculty of using rifle, sub-machine-gun, grenade, knife, etc, notwithstanding how the model soldier is armed. Thus, a figure in an action-pose throwing a hand-grenade for example, might actually throw it on Move 1, fire a submachine-gun on Move 2, and use a knife in a hand-to-hand situation in Move 3. It is recommended that figures be

the strength of the ambush, and he may continue into the ambush in a prepared condition, or move in a different direction. If he continues into the ambush a die is thrown to see whether he or the ambushers actually fire first. A derivation can be that when a scout sees an ambush he is able to fire a warning shot or blow a bugle, which can be heard by all figures on the table within 24 in (60cm). A die is thrown and score of 1, 2 or 3 means that no alarm is given; 4, 5 or 6 means that the scout has alerted his colleagues in the immediate vicinity. Even when so alerted, the column *must* continue into the ambush area but the ambushers themselves do not automatically fire first. Once the ambush has been detected, the ambushers may reveal themselves and retreat if they wish. If the ambush has only 6in (15cm) (or less) range of vision, i.e. around a corner of a track, the ambushers are as blind as the defenders and must dice for order of firing.

When an ambush takes place with the oncoming column being surprised, only half of them can fight the melée. If they kill more than they lose then the effects of the ambush are deemed to have worn off and – if there is another round of melee – they can fight with their full numbers. If they lose most men then they fight the second round with only half of their remaining numbers and, if again losing more then morale, rules must make them break and run.

based in threes, on irregularly-shaped pieces of card bearing say a rifleman, grenade-thrower and a submachine-gunner; heavy machine-guns, mortars, anti-tank weapons and flame-throwers should be mounted as specific teams.

In action, casualties can be determined by nominating a group's target; throwing an ordinary (ie 1 to 6) die; and a group is knocked-out and removed according to weapons used and die-score, thus–

Weapon	Die Score
Rifle; Sub-machine-gun	6
Light machine-gun; Grenade	5 or 6
Heavy machine-gun; Mortar; Artillery	4, 5 or 6

When actually coming into contact, opposing players each throw an ordinary die, multiplying the number of men in each group by the die score – airborne troops count $1\frac{1}{2}$ and ordinary infantry 1 point; troops defending hard-cover (house, wall, etc) counting double points value. The group with the lowest total withdraw their move-distance.

At Crete, the Germans employed ten troop-carrier groups, each composed of fifty JU52s and DFS230 gliders – the same combination used in their pioneer-effort of a year earlier when storming the fort of Eben Emael in Belgium. The glider carried two pilots and nine fully-equipped men; a carrier aircraft lifted 13 paratroopers – if that is scaled down for wargaming purposes it still results in a substantial number of aircraft whether they be actual or hypothetically considered on a chart. Far better when simulating para-drops to represent the entire air fleet by a single 'aircraft' which is itself simulated by a plastic model of a JU52, or perhaps a lidless cardboard box containing the flimsy paper shapes that symbolise the paratroopers, until replaced on the tabletop by actual model soldiers. Gliders, scaled-down in number, are simulated as in the article, Eban Emael.

With hindsight, the German invasion of Crete falls neatly into a pattern with the action at Maleme all important and German activities at Heraklion and Retimo merely containing local garrisons – when Maleme was captured the results of the other two areas were unimportant. On the other hand, Operation 'Market Garden' required an essential linking-up between airborne forces seeking to capture a series of bridges while a relieving force battled to contact them. Thus, it is more essential to re-fight 'Market Garden' as a campaign with an overall result.

As a means of getting a club or group off the ground (an apt turn of phrase in the circumstances!) this method of wargaming can hardly be bettered because it is cheap, easy to do, the battles are still familiar, and they have enough modern technology and derring-do about them to arouse interest and enthusiasm.

'Stand to the door! Green on! Go!'

War in the Sudan – illustration from Illustrated London News *1885 by Caton Woodville 'On the Road to Metammeh'*

DOWN AT THE WARGAMES CLUB No. 2

One of the most enthusiastic members of the Wargames Club was Billy Phillips, whose Dad had been a Regular Soldier in the Rifle Brigade with a lot of service in the Western Desert, France, Cyprus, Palestine and Borneo. For some of his earlier life Billy had lived in Army Quarters but now Mr Phillips was out in Civvy Street and they lived in a fair-sized house on the Estate that was a real military museum. When you went through the front door into the hall the first thing you saw was a huge framed reproduction of Beadle's painting 'The Rear Guard', depicting the 95th Rifles covering the retreat to Corunna in 1808/9 – you know it, they used it for the dust-cover of Bryant's book *Jackets of Green*, and the original is in the Greenjackets Museum at Winchester. On the walls were a fine assortment of weapons including a genuine Baker Rifle that must be worth a lot; and the house always echoes to recorded music of the Regimental Band playing Light Division tunes, like 'I'm 95' or 'Over the Hills and Far Away' and 'Lutzow's Wild Hunt' – it was really stirring! Billy was brought up a real little Rifleman, taken every year by his Old Man to Winchester for the Sounding of Retreat Ceremony, then spending the rest of the evening in the Sergeant's Mess, when they made a real fuss of little Billy! One year we all went to the Massed Band show at Wembley and were all very impressed when we discovered Billy actually knew Bugle-Major Colin Green, the little man with the fierce moustaches who strutted along at the head of the Band.

With all that background it was'nt surprising that when we all picked a regiment for a wargames campaign project, Billy chose the 95th Rifles. Mind you, he had opposition as another lad wanted them and it took a bit of pressure to persuade him to have the Connaught Rangers instead. When he heard of Billy's choice (not that he'd have dared make any other) his Old Man stumped up the cash to buy forty 25mm figures, including officers and buglers. Researching uniforms and equipment, Billy painted them with real devotion and when he marched them onto the wargames table we were all full of admiration as they looked great in dark green tunics with black facings, three rows of silver buttons, black leather equipment, stovepipe shako with green cords and tuft, and silver buglehorn badge. The N.C.O.'s had white stripes, and all were armed with Baker Rifles and long brass-hilted sword-bayonets. On the underside of each figure's base was a little label bearing the soldier's name – there were officers Harry Smith, Johnny Kincaid

and George Simmons; there was Harris, Surtees, Costello, Plunkett (the man who shot the French General on the bridge at Cacobellos) and the rest of them. They were commanded by a fine mounted figure labelled Colonel Beckwith and for good measure, Billy painted-up a realistic General 'Black Bob' Craufurd, who died in the breach at Ciudad Rodrigo in 1812.

Wargamers being what they are, there was a great deal of argument over the campaign rules and no one gave more trouble than Billy – claiming the 95th to be '. . . a highly mobile elite regiment . . .' he demanded special rates of movement to simulate their 140 paces-to-the-minute, higher morale-rating, and no penalties for fighting in open-order. He wanted battles framed so that they fulfilled their historical role of skirmishers ahead of the main line, harassing the French tirailleurs, holding rear-guard positions, and so on. You've got to admit that on the table they were a revelation, fighting furiously and never routing, perhaps because Billy had the most fantastic dice luck! On the other hand, maybe it was his unflinching confidence that got through to their little metal hearts, as he ordered them about the table by name, in time he knew each one without even having to turn up their bases to read their labels!

Then we were invited to go to London to take part in a big Waterloo wargame, and Billy had to wrestle with his conscience about some of them being at New Orleans rather than in Belgium on 18 June 1815, but eventually the 95th were brought along to take part. The hall was very crowded and the guy guarding the table at lunchtime never noticed the thief who helped himself to the entire 95th Rifles! Of course Billy went mad, wanting the Police called and everyone searched, using bad language when the organisers offered money to buy replacements; he had to go home without them, sitting silent and unconsolable and trudging off into the darkness, a shattered man, without even saying 'goodnight.'

For nearly a year he never came near the Club, rumour saying he had left home and gone into lodgings, breaking his Mum's heart. Then, late one Saturday night after a wargame, just as we were going for a drink, Billy rushed in and from a cardboard box took figure after figure and set the 95th up on the table! They were a battered, chipped and sorry-looking lot but the 95th alright, every man-jack of them, down to their original labels on the bottoms of their bases. Words falling over each other, Billy told how he had gone to this Hobby Exhibition (he said the Rangers were playing away and he'd nothing else to do) and there, in a demonstration wargames were his 95th Rifles, controlled by a scruffy Wally who claimed he bought them off a guy in London. When Billy claimed them, the Wally demanded proof of ownership – '. . . as if I didn't know my own figures!' snorted Billy: 'I identified them by name . . . in order of seniority of course . . . Colonel Beckwith, Harry Smith, Johnny Kincaid, Sergeant John Lowe . . . Riflemen Harris, Surtees, Costello, Leach, Howans, Jackson, Bugler Bill Green . . . without looking at their bases, of course!' Well, the Wally had no option but to hand them over, saying he didn't mind very much 'cos they never fought very well for him! 'I nearly hit him . . . of course they wouldn't . . . they didn't know him.'

He spent weeks repainting them, but when they paraded for battle again they weren't the 95th we remembered; they fell-back, wouldn't advance, then Billy threw a 1 and they all routed back to the base-line – we thought he was going to break into tears! He took them away and it was generally believed solo-wargamed with them, to get them reacquainted with his command and, as weeks passed, slowly they began to improve, became steadier. Then, one sunny afternoon in May when some of the Gang felt it was too nice to stay indoors

wargaming but got talked into fighting a Napoleonic Peninsular game, they came good again. Charlie our President, who had a real sense of military history, had planned the battle and in the light of what transpired, it was probably no coincidence that we were refighting the Peninsular War battle of Fuentes D'Onoro and Charlie had taken command of Captain Ramsay's Troop, Royal Horse Artillery. The game was designed to allow the guns, just as they did on that immortal day, to make a fighting retreat across the table, limbering and unlimbering as the riflemen fired and doubled back, fired and doubled as they held off the hordes of prancing French cavalry. Skilfully and realistically, Charlie and Billy fought their way across that Spanish plain – aided by some damned good dice-throwing – and the 95th were never better, they had regained all their former glory!

After the game, Billy said: 'I had no doubts . . . I told them that today the 5th of May is the 175th Anniversary of the Battle of Fuentes D'Onoro . . . and, knowing what they did then, how could the 95th be anything else but magnificent?' You had to admire his confidence, didn't you? It must have sent his old man mad when Billy preferred the Art College to joining the Green Jackets!

6
THE 'PURE' WAR –
Tanks in the Western Desert 1940–42

Probably more than any other campaign of modern times, the battles fought between the 8th Army and the Afrika Korps in Libya during 1940, 1941 and 1942 were relatively uncomplicated, being mainly affairs of movement between armoured forces with infantry in a subsidiary role. This Western Desert Campaign has many unique and glamorous aspects that make it attractive to the wargamer, yet few attempts are ever made to simulate it in miniature. Perhaps this is because, although seemingly easy to reproduce on the wargames table, it is a campaign that possesses built-in difficulties caused by extreme natural factors allied to the man-made mechanical complexities of armoured vehicles and ballistics.

Possessing a particular flavour and life, the Desert War held a fascination all its own. As a fighting arena the desert was perfect because it allowed as close a straight-out trial of strength as is possible on any battle front on earth. Seemingly difficult for either army to carry out a surprise attack because of the natural obstacles of the desert and because the preparation of mechanised forces for a big offensive on a large scale could not be concealed, an amazing degree of surprise was nevertheless attained, particularly during the first battles of the campaign and before the final British attack at Alamein.

War in the desert could be likened in many ways to war at sea with the reconnoitering armoured cars like scouting destroyers whilst the guns and tanks were the battleships and cruisers. The very language in which directives were coined had the same ring – the traditional naval order of 'Search out and destroy' was a reasonable maxim in the desert. Similarly, just as modern sea power could be destroyed if not adequately supported in the air so could the desert armies when one side had air superiority. Another resemblance to sea warfare lay in the manner in which movement, the key to everything, was restricted by the mine-fields which covered vast, often uncharted, areas.

There was nothing static about the Desert War. For nine-tenths of the time certain areas of ground were empty whilst at other times they were important. It was not ground that was wanted, both sides realised that there was no virtue in capturing great tracts of sand unless one could take it all – the more desert that was conquered, the more supply problems were aggravated and the more vulnerable communications became. None of the dilapidated Italian villages and the thousands of miles of desert

Model of German PzKpfw IV Tank, used in the Western Desert

73

around them were the slightest value to either the British or the Germans. In fact it would not have been particularly important to the British if the Germans had suddenly seized a thousand square miles of desert to the south: providing the British 8th Army was still in existence at Gazala, the Germans could not hope to hold this territory.

The aims of both sides was to get at the enemy, navigating across the desert with a compass just as at sea, and sweep him clear out of the area. Neither side came into the desert for loot or conquest but simply for battle.

A thing of fast movement and new tactics, this type of war had one main consideration – the balance of mobile striking power, a term which may be used to cover the combining of tanks, anti-tank guns and field artillery working together as a common force. Tactics were greatly influenced by geographical factors in that there were no defensive positions that could not be outflanked, except at El Alamein and at El Agheila. At both these places secure flanks were provided by narrow bottlenecks – between the Mediterranean and the Qattara Depression in the one case and the great Sand Sea in the other. Everywhere else there was an open desert flank causing constant anxiety to the commander who had not got superiority in mobile striking power.

Relatively few soldiers died in the Middle East because this type of open mechanical warfare tended to destroy machines rather than men. There was practically no trench warfare in the desert and once the protective armour was gone there was little the infantry could do except surrender, so the huge number of prisoners were nearly all unwounded. In the desert, once the enemy armoured forces had out-flanked them not even the finest infantry could hold fixed positions short of establishing them-selves behind a fortified perimeter such as Tobruk, possessing a water point and harbour to which supplies could be brought in. Otherwise, if they were not to be cut off, the infantry had to continue withdrawing until the enemy lost control of the open desert flank because of reduction of his superiority in armour and mobile forces through wear and tear and supply problems.

The desert permitted textbook-like application of armoured forces. It was soon obvious that the British principle of attaching tank forces to infantry units was inferior to the German concept of armoured tactics. To this must be coupled the acknowledge-ment that the German generals in the Desert were superior to the British generals in that they made fewer mistakes and did the right thing more often. Rommel was an abler general than any on the British side and the army he commanded was an abler force than the British army. The British did not seem to be able to marshal and drive their tanks as the Germans did – perhaps they were simply not trained so highly as the much-practised Germans, perhaps they did not possess the same innate feeling for armour. British gunners and drivers could be trained for their mechanical job so that many in the desert were equal or superior to their German opposites. But it took much longer and a differ-

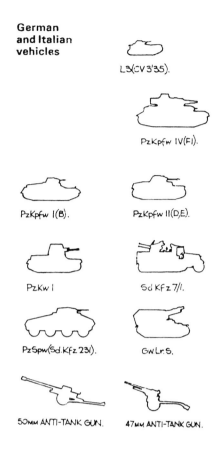

German and Italian vehicles

L3(CV 3'35).

PzKpfw IV(F1).

PzKpfw I(B).

PzKpfw II(D,E).

PzKw I

Sd KFz 7/I.

PzSpw(Sd.KFz 231).

GwLr.S.

50mm ANTI-TANK GUN.

47mm ANTI-TANK GUN.

A World War II armoured action in a Middle East setting. 1:300 scale figures of Richard Ellis Collection (Courtesy Old Town Studio, Swindon)

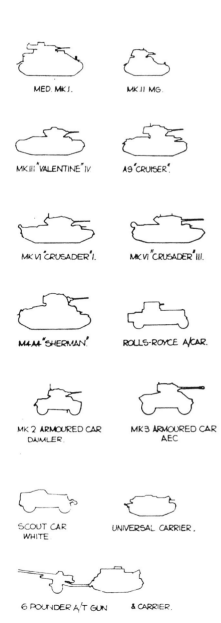

MED. MK I.

MK II MG.

MK III "VALENTINE" IV.

A9 "CRUISER".

MK VI "CRUSADER" I.

MK VI "CRUSADER" III.

M4A4 "SHERMAN".

ROLLS-ROYCE A/CAR.

MK 2 ARMOURED CAR
DAIMLER.

MK 3 ARMOURED CAR
AEC.

SCOUT CAR
WHITE.

UNIVERSAL CARRIER.

6 POUNDER A/T GUN

& CARRIER.

FORD WOT 6 TRUCK.

MORRIS COMMERCIAL TRUCK.

ent sort of brain or outlook was required to command a tank, whilst the man who can handle large groups of tanks was a much rarer bird who needed, among other things, considerable experience in actual warfare. The generals, the brigade commanders and the Intelligence Officers do not clearly or fully see the action in tank battles. It is upon the individual commanders of the tanks that the real responsibility falls. Without much direction from outside they fight an isolated and restricted battle, seeing only local incidents because the whole business was too quick and complicated and too obscured by dust and smoke. In the same way as a man saw only a limited sector so the battle itself was divided into a series of fast-moving incidents which would only be formed into a complete picture on a map when the battle was over. In the long run it was equipment and training that counted and the Germans were superior in both fields.

German control of armour was superb because they were a highly trained group of self-contained tank technicians controlled en masse as easily as a single vehicle. In action, tanks, anti-tank guns, recovery vehicles and petrol waggons, all supported by Stukas, went forward rapidly and successfully, often accompanied by senior officers. The co-operation between armour and the anti-tank gunners was exceptional, in fact the brilliant successes of the Afrika Korps depended upon three factors – the superior quality of their anti-tank guns, the systematic practice of the principle of the co-operation of all arms, and their tactical methods.

Rommel elaborated the basic German battle drill and so made possible an instant cohesion of small units of tanks, artillery, infantry and armoured cars. They fought as a team whereas the British, despite brilliance and doggedness, fought as individuals until they learned better.

Rommel is reported as saying:

'I don't care how many tanks the British have so long as they keep splitting them up the way they do. I shall continue to destroy them piecemeal.' (*African Triology* by Alan Moorhead (1944).)

Actually on most occasions the Afrika Korps moved so fast and decisively that they, rather than the generals, controlled the movement of British armour.

A highly flexible formation of all arms, the German Panzer Division, always relied on artillery, trained to co-operate with the Panzers, in attack or defence. Having experimented with their 88mm anti-aircraft gun as far back as the Spanish Civil War, the Germans employed it to shoot at tanks as well as aeroplanes. In November 1941 they had only 35 88mm guns in the Desert but in close co-operation with their tanks these guns did terrible execution amongst the British armour. In addition, the German high velocity 50mm anti-tank gun was far superior to the British 2pdr and batches of these guns always accompanied German armour into action. On the other hand the British restricted their 3.7in anti-aircraft guns to an anti-aircraft role and regarded

A model of an American General Grant tank, a welcome reinforcement to tank strength of the 8th Army 1942

the anti-tank gun as a defensive weapon. The powerful British artillery was never adequately used to eliminate the German anti-tank guns.

Knowing that gun power had to be balanced, the British reasoned that the 25pdr was primarily offensive, the 2pdr was defensive, and the 6pdr was offensive up to 2,000yd (1,800m). The 75mm gun with which the Grants and Shermans were equipped could destroy German armour and also their dug-in anti-tank guns. The Crusader tanks were fast and inconspicuous but had to close to effective range – either you went slap in with a Crusader or you kept out and did not dally in the danger area. Often the tactics were to hold the enemy with 25pdrs and 6pdrs in support groups whilst the tanks went round to the flank. The first tank wave penetrated as far as possible and the second wave came in behind and dealt with the fringe of the enemy armour. This was all right in theory but did not always work in practice.

The standard British tank tactical formation was the arrow-head which allowed each tank to give good fire support to its neighbours providing at the same time a fairly good defensive formation against anti-tank gunfire. Being undergunned and under-armoured, the British were forced to resort to tactics of ambush or rushing in as quickly and as close as possible, accepting losses in the hope of dealing a quick knock-out blow. Apart from the 25pdr field gun, the only way that the 2pdr anti-tank gun could cope was to dig in and fire at targets obliquely across the front to penetrate the thin side armour. The British possessed a debatable advantage in numerical superiority of their faster cruiser tanks. On the other hand their slow moving heavy infantry tanks were tied to the otherwise helpless infantry.

Gun out of Action!
In wargaming it is usual for casualties to be caused amongst artillerymen rather than to have the gun itself damaged or put out of action. An alternative method is to give the gun itself a points value so that when it is fired upon by enemy artillery, the casualty score can either be taken in gunners or deducted from the points value of the gun itself; decided by dice. A small chart can indicate that, after so many points have been lost, the gun has a wheel broken which must be replaced (taking up 2 game-moves), until finally the gun is permanently out of action because its points score has been eliminated. Allocate a points score of say 30 for a field gun and 20 for a horse gun, and when 50 percent of that total has been eliminated, then the gun is partially out of action i.e. for a wheel exchange etc. When the whole sum is eliminated the gun is completely out of action.

Armour on Wheels
The Armoured Car since 1912
Among the earliest recorded instances of motor-driven armoured vehicles in action was in 1912 when the Italians used a 4 ton Bianchi armoured car in their colonial activities in Libya, and also in the Balkans War. The first-ever British mobile armoured forces were Naval not Army under the command of belligerent and unconventional Commander C. R. Samson RN, who in September 1914 had a squadron of ten aircraft based on Dunkirk with the role of denying the French and Belgian coast to German aircraft and to attack Zeppelin bases in Germany. The Admiralty procured 100 Rolls Royce cars and formed them into squadrons to protect temporary airbases and to rescue any pilots forced down in territory overrun by the Germans. Soon, the cars dominated the country for miles around Dunkirk – their speed and mobility plus the firepower of their crews making them highly suitable for the open warfare of that time.

Frequently coming into conflict
continued

The favourite German attack formation was in a block five or six abreast in line astern, with the heaviest armour in the van, anti-tank guns on the flanks with a sprinkling of motorised infantry. The main force of motorised infantry were in the rear, covered by field artillery. In close support were the Stukas to deal with any obstinate British strongpoints. If repulsed, they would skilfully draw the British tanks on to their anti-tank guns – Rommel liked to get British armour to attack him when he would dispose his own armour behind carefully camouflaged anti-tank guns on which the British tanks were lured. The Germans liked to bring anti-tank guns boldly forward to the front of an armoured battle, leapfrogging them from one vantage point to another while their Panzers, stationary and hull-down if possible, provided protective fire. With the anti-tank guns in position and themselves given protective fire, the Panzers swept on again. The tactics worked well and the British tanks steadily sustained losses, had to give ground constantly and were rarely able to hold up the German advance.

Typical German tactics in the desert lured attacking British armour towards the 'soft' targets such as a troop or truck concentration. Then the British armour was engaged by well dug-in anti-tank guns on the flanks. PzKw III and IV tanks would then engage at long range, finally moving in to finish off the trapped British vehicles.

The British misunderstood the way the Germans handled their armoured formations. They did not appreciate that Rommel's Panzer groups worked on the principle that whereas tanks dealt primarily with the enemy's infantry and soft vehicles, the destruction of enemy tanks was mainly the job of the anti-tank guns, the weapons designed for just this purpose. The Germans did not commit themselves to tank versus tank battles as such. On the Alamein Line and outside Tobruk they avoided tank

action unless they greatly outnumbered the British.

So it can readily be seen that the Western Desert Campaigns are made-to-measure for the wargamer, simulated as an all-vehicle affair with infantry only being represented as carried in half-tracks or trucks. This obviates painting figures, one of the biggest chores of the hobby, while the miniature vehicles can be sprayed sand colour, for example, in a matter of seconds.

To give some idea as to what can be done in the way of simulating armoured warfare in the Desert let us imagine two armoured forces, one at each end of our table-top Desert – at least, it is thought that they are there but as they are not in view neither opponent has any idea of the whereabouts or size of the enemy force facing him. When the two forces are considered to be within vision-distance – when they are placed upon the table – a wargaming simulation of conditions peculiar to the Desert War will come into operation. The most effective factor is for both wargamers to fight their battle *sitting down*, one on either side of the wargames table, their forces being moved by other hands. This simple ploy at once introduces a factor of realism almost unparalleled in wargaming by diminishing the God-like status of the players as they tower over the field with each and every aspect and movement open to their eyes. In other words, it introduces the realistic and invariable situation on battlefields of lack of vision, and in doing so allows concealment and surprise, those most elusive aspects of tabletop wargaming. On the far side of a mildly undulating field, perhaps 6ft (2m) from your eyes, lays a force of enemy armour, each tank so small as to be able to sit comfortably upon a fingernail; adding to the problem, they are painted a camouflage blend of light brown and are standing upon a desert-sand coloured field! Another bonus for realism lays in that your orders to the hand moving your army – made verbally or in note form – stand a very good chance of

with enemy infantry and cavalry patrols, Samson had the inspiration of hanging boiler-plates to the sides of the cars to protect the crews. Later, the Rolls-Royces were provided with temporary armour by having steel plates fitted at the Chantiers de France steelworks at Dunkirk, arranged in the form of an open box body which enclosed the rear wheels; triangular-shaped boxes protected the front wheels, a small raised box protected the driver's head, and the engine and radiator were covered by flat plates. The armour covering the body of the car consisted of ¼in (6mm) thick double plates separated by 1in (25mm) board. The boiler-plate was vulnerable to German armour-piercing bullets, ½in (12mm) thick being required to withstand them, but such a thickness would have been so heavy as to reduce the speed of the cars greatly. Subsequently the armour was limited to ⅓in (8mm), the crews knowing that they had to keep the enemy at 500yd (460m) to be safe.

continued

*Dawn in the Western Desert 1941
(HMSO Publication 1944)*

Each car was armed with a
Maxim machine-gun on a
pedestal mounting, immediately
behind the driver's compartment.
So successful were these cars
that Rolls-Royce began fitting
proper armour plates to the cars
they were making for the
Services, extending it over the
top to give complete protection
and adding a revolving gun
turret. The first three turreted
Rolls-Royce armoured cars
landed in France in December
1914; they retained their classic
lines, and had a bevelled turret
mounting a single Maxim water-
cooled machine-gun. The cars
had electric lighting, twin rear
wheels and pneumatic tyres, and
detachable running-boards for
using crossing obstacles; their
armour was 0.30in (7.6mm) thick.
    At the same time as
Commander Samson's 'guerillas'
were motoring around in France
continued

being misunderstood so that your tanks do not do exactly what
you had in mind – thus is reproduced the vagaries of communication by tank-radio!

When the distance between the opposing forces closes, then
hopefully identification can be made by noting the shape of a
vehicle and hastily scanning your Silhouette Chart. This is an
adaptation of the Naval method of identifying unknown vessels
by reference to such sources as *Jane's Fighting Ships* as soon as
it is possible to discern the silhouette of the approaching vessel
and from that data identification can be made.

This singular method of fighting World War II tank battles
differs from methods commonly used in wargaming and allows
dividends to come to the seated wargamer who, having studied
the tactics of the day, employs them upon his tabletop desert. If
they worked in real life, then the methods of fighting employed
by Rommel should be successful in miniature, because wargame
tanks and soft-vehicles can be knocked-out just as easily by
simulated methods of tank and anti-tank gun fire. Methods of
doing this on the tabletop can be found in the author's books
Tank Battles in Miniature (published in 1973 and 1977 – now out
of print but obtainable from Public Libraries). Rules should be
so formulated as to represent the known superiority of German
tanks and their anti-tank guns; minefields can be pre-laid in the
manner described elsewhere in this book.

Both sides were plagued from the air, mostly by strafing
or low-level bombing and, in the case of the Germans, by the
practised use of their dive-bombers. In return, the Germans were
often on the sticky end of some sabotaging expedition made by
the Long Range Desert Group.

Ground scale must obviously be affected by the scale of the
models that are used but as it is so obvious that 1:285 scale
model tanks are the best for our purpose then a scale that suits
them will be discussed. It will no doubt arouse considerable controversy to disassociate ground scale from the actual scale of the
models but it simply would not work unless the wargamer has a
vast table-top arena at his disposal. For example, if the ground
scale of 1:285 (the same as the smaller suggested vehicles) is
used then a wargames table 6×6ft (1.8×1.8m) would only give a
frontage and depth of about 566×566yd (510×510m). Of course,
this is far too small an area to use because it is a ridiculously
close range for any gun such as the 88mm – although the British
2pdr might appreciate it! The ideal scale would seem to be 1in =
50 yards (1:1,800) which would mean that our 6×6ft (1.8×1.8m)
table would now become about 2 miles (3.2km) long and about
2 miles (3.2km) deep. Using a time scale of 1 minute and a
movement scale of one inch = 50 yards, then a vehicle travelling

at 1 mile per hour will move .58in
at 5 miles per hour will move 2.92in
at 10 miles per hour will move 5.84in
at 20 miles per hour will move 11.68in

Metric examples:
    at 2kph will move 1.8cm
    at 10kph will move 9.3cm
    at 15kph will move 13.9cm
    at 30kph will move 27.8cm,
while a plane travelling at 315 miles (500km) per hour would cover 184.8in (4.7m). This would mean that a tank travelling at 15 miles (24km) an hour (the average speed for German Mark III and Mark IV tanks) would move 8.8in (22.4cm) and an armoured car at 45 miles (72km) per hour would move 26.4in (67cm). On the subject of tank speeds, it must be appreciated that they might well move at two different rates, (a) battle speed and (b) road or 'getting somewhere' speed. In the desert circumstances these two may blend so that it is not unreasonable to work out tanks' moving distance (using the scales used above) at two-thirds of their road speed as a general rule. On the other hand, desert conditions were sometimes so good (on the salt pans) that tanks could move as fast or faster than on the road, whilst in sandy or rocky areas their rate of movement was reduced to a minimum. Perhaps this could be covered by having an overall rate of movement of, say, half their road speed with an 'emergency' rate of movement of three-quarters of their road speed (this approximates to the charge-move given to cavalry in Horse and Musket wargames).

Our ground scale not only affects movement of the vehicles but also the ranges of guns and, at 1inch = 50yd, 1,000yd (900m) range will equal 20in (51cm) and an 88mm gun firing at 2,500yd (2,250m) will have a range of 50in (125cm) on a tabletop terrain.

Perhaps this might not be the appropriate place in this book to mention the fact but it is better to include it here rather than inadvertently fail to include it at all. One of the most powerful factors in wargaming during all periods of history lies in the morale of the troops concerned. Troops with a high standard of morale can carry on longer and show better results than their counterparts whose morale is shaken by exhaustion, losses or other discouraging factors. No suggestions have been made in this book regarding these factors but it seems absolutely imperative that the morale state of a tank crew or tank crews should come into question when they have been fired upon. Wargamers can work out suitable morale rules to cover these contingencies and might even feel that the morale of the Allies and the Afrika Korps could well vary according to the state of the campaign. For instance, the morale of British tank crews was extremely low at varying times because of their knowledge that they could be penetrated by German tank guns at far greater ranges than that at which they could ever successfully retaliate. Then there is the impact of such defeats as the chase back to Alamein (although, with true British perversity, it would seem that things were not as black as they might have been painted at the time). Similarly, the Afrika Korps might have a lowering of morale as a result of the continued chase back from Alamein into Tunisia.

and Belgium, the Russians on their front were using armoured cars built on a British Austin chassis; driving only on the rear wheels, their cross-country performance was limited. The Germans also used armoured cars – Daimlers and Bussings – driving on all-four wheels, but little is known of their activities. Armoured Car Squadrons played an active part in support of British infantry at Ypres, operating as the Dunkirk Armoured Car Force; while the roads remained open, they invariably dominated the actions in which they fought. But when the war began to bog down in trench warfare, their mobility became increasingly restricted, and by the end of 1914 they had all been shipped back to England, and eventually went to the Middle East, where, independent of roads, they were highly effective.

The wargamer can set-up interesting small-scale actions recreating this fluid period of World War I, using armoured cars – possibly scratch-built around a model vintage car – combined with infantry, cavalry, motor-cyclists, machine-guns and artillery. There are obvious facilities for surprise, concealment, ambush and numerous other aspects of warfare that make for a good tabletop wargame.

DOWN AT THE WARGAMES CLUB No. 3

Most of the time we all got on very well down at the Club – there was always a lot of kidding and winding-up but it wasn't taken seriously and we were really all good mates. It carried on like that until Sharon joined us, the first girl we'd had in the Gang – not because we were male chauvinist pigs but simply because none of the opposite sex ever came along – but the experience was enough to cause us to hope she will be the first and the last. In the beginning we heard of her through Sam Russ raving about this girl he'd met – 'I was having a quiet pint in the Dolphin when this couple sat at my table . . . she was a real cracker but the guy with her was a dreary Wally who seemed to be boring her to tears and after a while she really let him have it, saying:

". . . I don't know why I'm wasting my time with you when I could be at home reading the third volume of Oman the library got for me today!"

Sam repeated: 'The third volume of Oman . . . that's what she said . . . I couldn't believe my ears . . . so I leaned across and asked her if it was Oman's *History of the Peninsular War* she was talking about?'

'The Wally pushed his face into mine and told me to mind my own business . . . but the girl told him to keep out of it and looked me up and down:

"Of course, it was, although I suppose you could have thought I was talking of his other books, say *The Art of War in the Middle Ages* or *War in the 16th Century* . . . I've read those, too." '

Sam took a deep breath: 'You're telling me you've read books like that . . . about history and wars?' The girl's eyes flashed in what we later got to recognise as a danger signal: 'And why shouldn't I? Those books . . . that sort of writing isn't for men only, you know! Of course I've read them although my main interest is the Peninsular war.'

'What else have you read on that?' asked Sam.

'Oh most of them . . . Jac Weller, Michael Glover, David Chandler, Rogers, Riflemen Harris, Kincaid, Surtees, Costello, Harry Smith, Julian Rathbone . . . and Napier too . . . but not all of his – only the single volume edition.

Sam had never met a girl like her and they chatted merrily away until the Wally got fed-up and went. Of course he had to bring her down to the shop, to show off this quite exceptional member of the fair sex whom he introduced all round as 'Sharon' and basked in the reflected glory of the astonishment and respect she aroused by knowing as much as any of us about the Peninsular – even Billy Wright hadn't read Napier! It wasn't long before she made it quite clear she wasn't going to be regarded as a pet parrot saying its piece; she stood there looking around at us and, as breezily as you like, said:

'Sam tells me you fight wargames here. I've never had the chance to do anything like that before . . . when can I come and fight a Peninsular wargame?'

We looked at each other and you could read it in their faces that they could see she would bring a bit of glamour to that dreary school-hall. 'It's alright with us . . . if Charlie agrees (the Club President).' Well, Charlie agreed, saying: 'Perhaps it'll make some of you watch your language when the dice don't fall right!' So we set about preparing a fairly straightforward battle, using an example from history and we had a full house next Sunday with most of the Gang more respectable than ever before.

Sharon astonished us straightaway by recognising the battle as soon as she heard the narrative: 'That's Maida . . . 4th of July 1806 . . . when Stuart beat Reynier . . . yes, that's a nice little battle with the red-coated Swiss being mistaken for Watteville's men . . . and Ross coming ashore with the 20th Foot and winning the day . . . yes, I'm going to enjoy this!' We stood openmouthed with Sam Russ capering around her as though she's just won an Olympic gold medal. At first we were all helpful and chivalrous, giving her the benefit of the doubt and acting like perfect gentlemen, then she got the hang of the rules and began knocking hell out of Toby Role's force. Mistakenly he carried on treating her with exaggerated courtesy long after most of us were desperately fighting for our lives! Oh yes, before the game was halfway through we had all abandoned that flippant flirting style, but Sharon had got the bit between her teeth and by the end of the afternoon had done a better job than even Stuart did in 1806! It might have helped if we could have got the odd curse or swearword in, but whenever one of us opened his mouth to do so, Charlie glared a warning – and it was his Club, after all, wasn't it?

From then on she became a regular Sunday player, coming each week with Sam, although he was getting a bit fed-up with Sharon and some said she only kept in with him to be able to carry on wargaming. Whatever you say, women aren't like men, are they – when a guy wins a wargame or does something noteworthy he doesn't keep crowing about it, but Sharon did because she was a real Woman's Lib type and every wargame was a Battle of the Sexes to her. Then she sprang her bombshell – she was going to paint-up a regiment of 25mm Amazons – women soldiers – and use them in our wargames! We kept our spirits up saying no maker did them, but Charlie said he'd look around although he didn't please Sharon when he recalled how Mike Blake of Individual Skirmish Wargames had once made a bevy of Western Saloon girls out of a box of Airfix 1:32nd footballers.

The eyes flashed dangerously: 'I'm talking of soldiers, women warriors who could beat most armies they encountered . . . until the controlled volleys of French repeater rifles defeated them in the 1890s.' And she went off full blast about this Corps of Amazons in Dahomey, a West African Kingdom, formed in bands of 400 with female officers, armed with muskets, rifles, bows and arrows, spears, machetes and swords, each unit designated by flags, drums and ceremonial umbrellas; they wore a loosely slit wide skirt and a cartridge-belt over their bare chests. That made some of the younger lads snigger! The local Model Shop produced female Fantasy warriors from Asgard and then Citadel's women-warriors, but she didn't like them so in the end to keep her happy, Sam converted some Ancient Egyptians or Hittities, gave them muskets and made bare breasts with tiny blobs of solder – when Sharon said some of them were '. . . unbalanced up top' we all said they'd look alright when they were painted.

She did a good paint-job on them, and those damned Amazons turned up in every wargame we fought, although at first the younger lads didn't like firing and meleeing with them – 'Don't seem right . . . with them being wimmin, does it?' But them 'wimmin' chased us all over the table and Sharon got more and more cocky, crowing when she won and the few losing occasions, accusing us of being chauvinist pigs. The numbers began dropping off on Sundays and we were at our wit's end wondering what to do, when suddenly Sharon stopped coming! After three peaceful weeks, we had all cheered up, even Sam Russ; then Fred walked in waving the evening paper: 'See this picture of Sharon in the *Recorder*?' It was on the back page, the sports page – there she was, dressed in football gear – she'd formed a woman's football team – called The Amazons of course – and was bitterly complaining because the local football association wouldn't let them play in the men's league! Understandable, you can't kick a girl up in the air.

We haven't had any girls in the Gang since Sharon, and most prefer it that way, although we all agree that if there was has got to be another sex, we'd as soon it was women as anything else!

7
GREAT-GRANDFATHER'S WAR

A gentleman in khaki

An ideal conflict for tabletop reconstruction, the Boer War of 1899–1902 has much to commend it – lots of movement; relatively small forces; great scope for cavalry; not enough artillery to take over the game; and well known battles, copiously documented.

The Second Boer War was unique in the area over which it was fought and the state of technology that provided the weapons. Earlier, without accurate artillery or automatic weapons, the Boers would have stood no chance in an era of smoke-enshrouded short-range slower musketry; ten years later vast improvements in field communications, plus the advent of aerial reconnaissance and armoured cars, would have made such a war impossible.

From sixteen to sixty, each Boer was a member of a local militia, grouped into commando units of various sizes to form an army made up entirely of mounted infantry capable of bewildering mobility. The Boer style of life as farmers and hunters taught them to be skilful with the rifle, and from boyhood each man's acquaintance with cover and terrain produced an irregular firepower capability from concealed positions capable of causing devastating casualties to close-order formations. They were armed with the same modern and efficient Mauser repeating rifles that had enabled the Spaniards to cause such heavy casualties to the Americans in Cuba. And they had purchased from Europe about a hundred of the latest Krupp field guns, which they handled with the same accuracy and efficiency as they did their Mauser rifles. However, both as individuals and collectively, the Boers lacked disciplinary control and except for a gifted few the majority of their leaders had little concept of tactics or strategy.

In the early days of the war first-class British infantry, in formations resembling that of the Imperial Guard at Waterloo, were launched in hopeless frontal assaults against the concentrated power of modern rifles competently handled by superb marksmen in carefully prepared positions. Although numbers were small by Continental standards, proportionate losses were very heavy. It took some time for the generals to realise the futility of sending men in close order over open ground against hidden riflemen, but then the British infantry began to advance in extended order in short rushes, sometimes to be successful against even the heaviest fire. Soon it was discovered that the

Boers, typically for partially trained Irregulars, were very sensitive to outflanking movements.

The American Civil War had caused British military thought to develop along the dangerously anachronistic line that trenches and earthworks were only of value to troops with insufficient training and experience to fight against Regular soldiers in open country, and that when first-class Regular infantry entrenched themselves their subsequent immobilisation produced a defensive attitude that lowered morale and blighted their aggressive spirit.

But had the untrained, irregular Boer farmers emerged from their entrenchments and advanced en masse across open country it would certainly have been the shortest cut to a Boer defeat. The Boers did not refuse to fight it out through lack of aggression but because they were fully aware of the potentialities and limitations of both their own troops and of the British. Skilfully, they husbanded their slender manpower as they made rapid adjustment to battlefield situations, such as quickly learning that soft earth cover (ie trenches) was far more satisfactory than hiding amid rocks and kopjes where bursting shells caused destructive rock splinters to glance in every direction. The Boer tactics of digging themselves in on features of great natural strength invulnerable to frontal assault positively invited outflanking movements by cavalry and horse artillery. But the Boers were

Boer War 1898–1902. The dreaded Vickers/Maxi 1pdr gun in a wargames setting

Simulating Modern Artillery Fire on the Wargames Table

Modern guns are too long-ranged to allow them to be placed on the wargames table with the enemy upon whom they are firing. Artillery fire is simulated by 'off-table map-shoots', which are carried out by marking the guns on a map and then concealing a figure (an observer) on the wargames table to lay the guns on to the targets he can see. To register a nominated aiming point, the observer must make a dice score of 5 or 6, which means that a hit is registered on that point (marked by a counter). Three such aiming points may be held at any one time. The observer may bring fire on to them unless the gun he is spotting for has moved (when they must be completely re-registered). The aiming points are also lost if an observer is killed. Guns may extend their target area by the observer nominating an aiming point and then altering the position of the 'Burst Pattern', providing a part of the pattern is still touching that point. A 'Burst Pattern' is a 6in (15cm) square of transparent plastic with four 2in (5cm) diameter circles numbered 1, 2, 3 and 4; the pattern bears a painted arrow which must point directly towards the firing gun. To simulate a shell burst, a pin is pushed through a centre hole in the 'Burst Pattern' into the 'hit' counter; the pattern laid, arrow directed towards gun, and a die score of 1 to 4 indicates the destruction of anything in that specifically numbered circle. A die score of 5 or 6 indicates the
continued

no fools, and, aware that such a position once enveloped was doomed, they were very alert to flanking movements that might turn their main position and on the slightest suspicion of such an occurrence they would slip away at night to another equally strong defensive site. This was a simple matter for the Boer farmer on his hardy little horse with well-worn saddle and bridle, carrying a blanket and a pair of bags for rations and spare cartridges with his rifle and bandolier around his shoulders.

It was soon discovered that not only was the fire of the Boer Krupp guns devastatingly effective but that it could also outrange the British 15pdr howitzers. The Boers had four large Creusot guns (nicknamed 'Long Toms'), capable of hurling a 96lb (44kg) shell a distance of 4 miles (6km), that were manoeuvred rapidly and competently by their gunners. But the weapon which the British soldiers dreaded more than any other was the 'Pom-Pom', a Vickers-Maxim gun that fired its little 1pdr shells in bursts of about twenty to wipe out complete gun crews or destroy carelessly grouped bodies of infantry simulated by placing a 2in (5cm) square of transparent plastic over the point of impact and throwing a die for each figure covered by the plastic, with a score of 1 or 2 indicating a casualty. To supplement the inadequate British artillery, long-barrelled quick-firing 4.7in 12pdr guns were brought ashore from warships and, mounted on land carriages, were handled by naval gunners. (These were the guns made by Britain's Ltd and immortalised in H. G. Wells' book *Little Wars*.) They had the range of the 'Long Toms' but their shells were too light to do much damage to a gun, although the crews could be harried into changing their position.

The rigid drills and customs of the Royal Artillery brought the guns into action neatly spaced in rows of six, making them extremely difficult to conceal. Then, though the guns were using smokeless powder, the still air and sun of South Africa caused a firing gun to generate a faint haze and after a few minutes of rapid fire six closely concentrated guns were covered by enough of a haze to betray and pinpoint their position. The Boers, with only a hundred guns at their disposal, dispersed them singly, carefully concealed along their whole front so that the slight haze created by a single gun was almost impossible to detect, making the British artillery fire at random. Battery commanders invariably sought to position their six guns on hard ground, but this allowed Boer high-explosive shells to burst with maximum effect, greatly magnifying the casualties caused by a single shell.

Wargaming Aspects

Generally speaking, these wargames will follow a pattern of small numbers of highly mobile Boers strongly positioned behind natural defences and in earthworks being attacked frontally by British infantry in close order. Should the attack show signs of success, then the Boers evacuate to another similar position when the whole business starts over again. When re-fighting the later stages of the war there is scope for mounted Boer attacks on convoys, night attacks on blockhouses and guerilla tactics. The more technically minded wargamer can scratch-build an armoured train and re-fight such actions as Chieveley, when Winston Churchill was captured.

shell has buried itself in the ground without exploding.

Artillery firing without an observer requires the map of the table to be divided into a grid of 12×12in (30×30cm) squares, each of them being again gridded into six 2in (5cm) squares, numbered 1–6. The firer nominates the large square, and dice thrown indicates the small square into the centre of which the 'Burst Pattern' pin is pushed. A further dice throw reveals the point of the shell burst.

Fire plans may be fired but must be initially written showing target area, number of guns, number of rounds and whether creeping or laddering up and down, etc. This firing plan should *continued*

The earlier victories of the Boers were largely due to a prudent habit of keeping out of sight (F. J. Waugh)

Yeomanry Cavalry, Boer War 1989–1902

be given to the umpire and once started must be rigidly observed except when stopped (not altered) by a staff officer gaining personal or wireless contact with each battery.

When fighting modern war games it must be remembered that artillery would not be so far forward as to actually be on the battlefield with the infantry and the tanks. To cope with this situation, 'off-table' shoots must be arranged, with each side having a certain number of guns marked on a map which is scaled to the wargames table. On the actual battlefield one observer per gun is concealed, his weapon being able to fire only on targets which he can actually see. In an effort at realism, one wargamer led pieces of cotton from each observer off the table to represent telephone wires, when one of these wires was cut by shell-fire or tank tracks, then the communication line between gun and observer was out of action and the gun could only fire blindly.

In modern war, fire-power is *continued*

Authenticity demands that the wargamer considers the varying qualities of the warring groups and in the Boer War the disparity between Boer and Briton was both marked and fascinating. The Boer was a better marksman than the British soldier and rules should allow him greater casualty-causing potential plus the ability to pick off officers and leaders. Although he had fewer guns, his artillery was more accurate and could outrange the British guns.

It is imperative to simulate the high mobility of the Boers, who moved everywhere on horseback, taking up defensive positions with their concealed horses close at hand. This can be represented by having three figures for each man – one mounted, one dismounted and a riderless horse. As their defensive style of warfare makes them extremely powerful, it will not be necessary to purchase large numbers of Boers, so that this three-for-one system is economically possible.

Conversely, British cavalry were slow and relatively useless except when fortunate enough to catch dismounted Boers in the open. The British infantry, until the later stages of the war, attacked and generally operated in close formation, making themselves ideal targets for the sharpshooting Boers. Casualty effects must be raised for the British in comparison with those for the Boers, who fought in scattered, open-order groups. Relatively few in number, the British mounted infantry were far more effective and could be given a fighting ability not far short of the Boers themselves.

As in most wargaming periods, morale is all-important and perhaps more than anything else is a means of breathing life into inanimate metal or plastic figures. The British, trained and stolid Regulars, should be classified so that they remain first-class until their casualties are relatively high; then there can be a short second-class period when they will remain motionless, holding their ground but incapable of going forward and finally after heavy casualties, they will go quickly (probably in rout) and will not return. The Boers, although stolid men of Dutch descent, display morale qualities that fluctuated from the peaks to the depths. This can be represented by rules that allow them to be unaffected until they have received high casualties or are threatened by outflanking movements, when their morale-rating will immediately drop to low and they will run for their horses and be off.

Here is a typical Boer War battle on which to cut your teeth – THE BATTLE OF MODDER RIVER fought on 28 November 1899. This battle was fought during the Second Boer War between the British 1st Division, commanded by General Lord Methuen, and Transvaal and Orange Free State Boer commandos under Generals Cronje and De la Rey.

Methuen's column, moving to the relief of Kimberley, had to cross the Modder River, which, like many South African rivers, ran through a miniature canyon about 30ft (9m) below the level of the veld, where a long line of bushes and trees marked its east

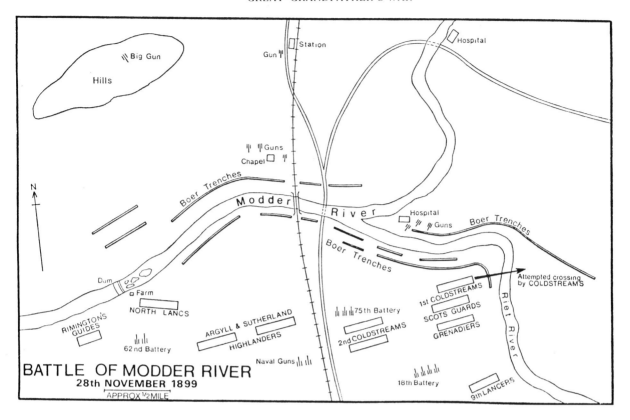

BATTLE OF MODDER RIVER
28th NOVEMBER 1899
APPROX ½ MILE

to west course. Methuen had no map of the area nor any details of the surrounding country, and discounted information about a weakly held drift lower down the river. He did not expect much opposition, as he was convinced that only about 400 Boers had been left behind to delay his force.

Anticipating the usual British frontal attack, the Boers had constructed elaborate defences: entrenchments masked by shrubs and brushwood stretched for 5 miles (8km) along both banks giving Boers on the southern bank a clear field of fire across the coverless veld, which sloped gently towards them. Their smokeless powder offered no visible target, so that the British artillery would almost certainly range on buildings on the north bank, where 7 field guns were positioned. There was also a heavy gun (probably a 100pdr) on high ground to the rear. Numerous artillery emplacements were constructed to delude the British into thinking they had silenced the guns when they had, in fact, been moved from one place to another. White-washed stones placed on the veld gave exact ranges both to the Boer gunners and riflemen. On the Tewee Rivier, a tongue of land between the Riet and the Modder, they dug-in a 'pom-pom' and two field guns, and they also positioned several Maxims and machine guns along their front.

In short lengths, the trenches were irregularly aligned, with their parapets concealed by rocks and bushes. Each held about six men. The farmhouses and buildings were converted into

often aimed more at persuading your opponents to keep his head down than at destroying targets. A useful rule is one which allows the battery to 'pin down' the enemy in an area say one foot square. Ranging-on is more difficult than maintaining the barrage and the odds should allow for this, similarly they should increase with range and the effectiveness of observation. All the enemy within the box are considered pinned down and may not move or fire while the barrage lasts. If they do so, or wish to move through the area, they lose a percentage, which varies with the type of troops, soft skin transport or armour. The percentage of loss can be increased by adding to the number of guns firing into the barrage area.

strongpoints. The Transvaalers were on the left of the position while the Orange Free Staters were on the right; it is thought that De La Rey positioned the latter with their backs to the river so that they would find it harder to break off the fight than they had done in their previous two battles.

About 6.30am the British 18th and 75th Field Batteries unlimbered on the right and opened fire at a range of 4,500yd (4,000m). Boer guns replied with a flash and a faint film of blue-white smoke instantly dissolving in the air. This long-range firing continued for some time, until the Boer guns ceased firing to give the impression that they had been silenced, and that their small rearguard was falling back. At that time the 9th Lancers, patrolling about a mile from the river, withdrew under a sharp fire to the extreme right of the line, where they took no further part in the battle.

At the regulation five paces interval, the Guards moved majestically down the slope, Scots Guards on the right, 3rd Grenadiers and 2nd Coldstream echeloned to the left, and 1st Coldstream in the rear. Descending the smooth grassy slope leading gently down to the river, they got to within 800yd (720m) of the enemy's trenches when suddenly from along the whole extent of the Boer front came musketry interspersed with Maxim fire. The advancing British fell flat to the ground, their slightest movement attracting a storm of bullets.

The Battle of Modder River, as seen from the Guards' Lines

The 1st Coldstream Guards, extending to the right to cover and support the Scots Guards, found to their surprise that they were halted by the Riet River running south to north, a fact completely unknown to the mapless Methuen. Nor was he aware that a few hundred yards back was Bosman's Drift, where a crossing could have been made in strength, taking the entire Boer position on the Modder in reverse. A small party of the Coldstream Guards managed to cross the river, but at a place that was obviously impossible for large numbers of men. The battalion dug themselves in so that the entire Brigade of Guards were out of the battle, and it was not yet 8am. With the halting of the attack, the artillery, in positions less than 1,300yd (1,200m) from the Boer trenches, spent the day covering the north bank with shrapnel.

Boer Artillery at Modder River

Far away to the left the 9th Brigade pressed forward, Northumberland Fusiliers on the right, King's Own Yorkshire Light Infantry in the centre, North Lancashires on the left, and the Argyll and Sutherland Highlanders in support of the right and centre. The KOYLI and Lancashires stormed a farmhouse in a kraal just to the south of the dam, getting within charging distance of the buildings by moving along an unseen and unprotected shallow ditch running down to a small clump of trees by the river bank. The Boers in the farm and in the trenches on either side were unable to fire on them for fear of hitting each other. Recklessly leading an attempt to cross the river, Lord Methuen himself was wounded and compelled to hand over his command to Brigadier Colville. Led by General Pole-Carew, the KOYLI, one by one, clambered across the river on the dam under a heavy fire, until 400 men had formed up on the other side; then they pushed along the north bank hoping to take the enemy in flank. Unfortunately, they were mistaken for Boers and fired upon by their own artillery, which compelled them to fall back. However, their appearance on the north bank so alarmed the lukewarm Orange Free Staters that at 2pm, a large number of them mounted and rode off; at 4pm there was something resembling a general stampede as the Boers retired along the deep river bed out of sight of the British, who did not realise what was occurring.

Earlier in the afternoon the 62nd Battery, after dashing 62 miles (99km) in 28 hours, arrived and immediately went into action, together with the rest of the artillery, as the centre of the battle moved towards the left of the position. An attempt to resume the Guards Brigade attack failed in the face of heavy fire.

In the early evening, after 8 hours of continual fighting, the remainder of the enemy retreated, unknown to the British, leaving behind their guns and many of their wounded. Later that night they mustered up courage to return and collect them. The British bivouacked on the field where they had fought and next morning made an unopposed crossing of the river to take the abandoned enemy position.

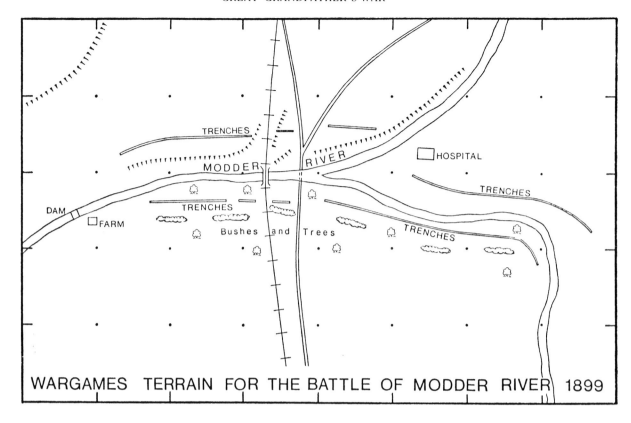

WARGAMES TERRAIN FOR THE BATTLE OF MODDER RIVER 1899

British casualties were 466 men killed, 20 officers and 393 men wounded, and the Boers lost 60 killed and 300 wounded, in what was a military farewell to the nineteenth century, in that the bewildered British soldier found himself severely punished through being tactically untrained to tackle a dug-in enemy possessing a high rate of firepower.

As the Boers have numerical parity with the British, plus superb cover, how can the British enter this battle with even the slightest chance of success? Here is a selection of Military Possibilities that might affect the course and eventual result of the action:

1. The OFS commandos are given a lower morale standard than their Transvaal comrades so that, under stress, they break and retire.
2. The British force can discover the existence of Bosman's Drift lower down the Modder River and cross there, turning the Boer flank. However, while this move may provide an interesting wargame, it takes all reality from the battle.
3. Pole-Carew's success on the left flank can be exploited.
4. The Coldstreams' abortive crossing on the right could be made a successful venture, again turning the Boer position.
5. The Boers' (particularly the Free Staters') lack of experi-

ence and dislike of being on the receiving end of artillery can be reflected by morale penalties when under fire, so that they might break at any point, particularly when the heavy guns of the 62nd Battery arrive.

6. The Boers' defensive dispositions were all made in anticipation of a British frontal attack, and De La Rey had early nervous moments when the Guards moved towards the flank before veering centrewards. A flank attack would have caused De La Rey to alter his dispositions completely, under artillery fire, or else withdraw. Again, however, this detracts from the realistic reconstruction of the battle.

Rating of Commanders and Observations

De La Rey stood out in comparison with the other leaders but to make him an 'above average' commander and Cronje 'average' is to overweigh the Boers' already strong hand. Therefore, both De La Rey and Cronje should be rated as 'average', and Methuen 'below average', but Pole-Carew, on the British left, could be rated as 'average'. Or it might be more realistic to rate De La Rey and Pole-Carew as 'above average', and Cronje and Methuen 'average'.

In spite of their inability to move forward, the British Regular soldiers cannot be considered to be low in morale or in fighting ability, if only because they possessed training and discipline superior to that of the Boers. The penalty for a low morale rating for the British soldier could be for them to 'go to ground' when they come under heavy Boer fire, when they have three choices:

A British infantryman in marching order

1. Once down they take no casualties but cannot fire back.
2. Remaining on the ground they can return fire but take half casualties.
3. They can fire on their feet and move their full distance, or they can move and fire with half effect, or they can charge-move and make contact with the entrenched Boers.

The condition governing each choice rests upon their state of morale under fire, decided by throwing a dice for each group and deducting 1 from it (a) for any losses, (b) if they come under fire from a group of men approximately twice their strength, (c) if they are under artillery fire, or (d) if they are under fire from flank or rear.

To be able to rise to their feet and move or fire, a group needs to total 3; to stay down and return fire the group needs to total 2; to stay down without casualties and without returning fire is their normal basic reaction, and is what they will do if unable to make the required totals.

The Transvaalers' morale will benefit from being behind cover but they should be rated as equal in morale and fighting ability to the British Regulars in fire-fights but slightly lower in melees because of their fear of the bayonet. Having shown a disinclination to fight at the early battles of Graspan and Belmont, the Orange Free Staters are regarded with some contempt by the Transvaalers, and it might be a reasonable Military Possibility to insert a built-in distrust in the game rules so that their actions are not well coordinated. The lower quality of the Free State Troops should be reflected in their morale and fighting ability.

Having received a grim lesson from British Lancers at Elandslaagte, the Boers greatly feared the lance, yet the 9th Lancers played no part in the Battle of Modder River. A Military Possibility could permit them to move from the right flank if desired.

The British artillery's firing on its own men on the left can be controlled by Chance Cards or, as the firing will be 'off-table' map-shooting, a certain allowance can be accepted for inaccuracy of aim. If it should happen that the British soldiers are hit, their morale must immediately be checked, with a 'distraction' factor to allow for their being fired on by their own side.

One realistic way of refighting this battle is to give the wargamer representing Methuen only the details that were known to the British commander in 1899 – 400 Boers forming a rearguard – only to discover, as did Methuen, that the situation is very different. As Methuen had no map of the area and discounted local information, the wargamer playing Methuen must be told that some or all of the information may be false; he has no alternative, therefore, but to make his plans and dispositions on a sketch-map conforming to his own vague idea of the terrain and what he is told that he can see. (This must be done before he sees the tabletop terrain.) The Boers will mark their dispositions on their own accurate map, and the troops from each side will be shown when action demands. Any casualties caused by artillery fire will be deducted on paper, the map being marked accordingly by the Boer commander, and morale checked whenever necessary. He will move on paper as if on the actual table, but any men breaking cover (such as in rout or in changing their position) will have to be revealed on the table.

Construction of Terrain

The Modder River will bisect its width just above centre, thus allowing ample space for the Boer positions in front of the river and for the British main area of action. There must be adequate space on the British left flank for the attack on the farm and the subsequent crossing of the river by the dam, and also sufficient on the right flank in case Military Possibility allows the Coldstream Guards to cross at the shallow point. The ground above the river should be slightly raised, so that the Boers have a field of fire over the heads of their men in the trenches forward of the river. The heavy gun can either fire from the hill at the top left-hand corner of the table or 'off the table' as a 'map-shoot'.

The battle might, in fact, be fought in two parts, as the British front was three miles long, and the soldiers at one end had no idea what was happening at the other.

Opposite:

Cheating a bit! Not 30 Years War, but an English Civil War siege, not differing in type of soldiers, weapons or style of fighting (25mm ECW Figures by Wargames Foundry, designed by Alan and Michael Perry, Photo by Duncan Macfarlane, Wargames Illustrated*)*

WARGAMING THE THIRTY YEARS WAR

The Thirty Years War 1618-1648 saw a dramatic transition in weaponry, ending the superiority of the fully-armoured cuirassier, armed with a long lance, who found he could not withstand the hails of fire poured on him by the musketeers protecting the great square-formations of pikemen. So the horsemen discarded all but 'back-and-breast' protection cuirass, to lighten the burden borne by his heavy horse; then armed himself with two to four pistols and employed the 'Caracole' pistol-fire attack. It was a sort of 'dance of death' as one wave of horsemen trotted in to fire at point-blank range, wheeled to the rear to re-load, while another wave 'danced' forward in the caracole, fire their pistols and wheeled away, to be replaced by the original first-wave, and so on. When it worked, it caused a constant hail of pistol fire to erupt on the pike-square until falling men caused a breach through which horsemen could pour to cut down the hapless defenders.

The caracole method of cavalry attack can be simulated on the wargames table by having the horsemen 'formally' move up on successive lines to the enemy, fire their pistols at close range, then wheel back to re-load and re-form their ranks. This allows interesting manoeuvring as enemy cavalry attempt to out-flank them and to counter-attack. When 'caracoling' a formation of pikemen, the following rules can apply:

Range of pistols 6″ – Throw one dice per man. 5, 6 is needed for a hit. (deduct 1 from dice if target is behind hard cover or wearing armour or if firer is moving when he fires.)

Saving-throws 5 or 6 infantry (+1 if behind cover, 4, 5, 6 cavalry.

A cavalryman can:

a) Move 12″ (15″ charge move) and fire 2 pistols. He is then completely unloaded until he halts and devotes a move to reloading.

b) Move 12″ (15″ charge move) and fire 1 pistol. he then has 1 loaded pistol remaining.

c) Deduct 6″ pistol loading from his move, moving 6″ + pistol loaded or nil +2 pistols loaded. He cannot charge move in the same as he reloads.

Opposite:
Another 17th century wargame –
Montrose against the Covenantors
in 1645. Figures and terrain, the
Chris Scott Collection. (Photo by
Richard Ellis)

A Push of Pikes

When simulating the shock-tactics employed by Gustavus's heavy Swedish cavalry – or indeed any form of cavalry attack other than caracoling – the rules must include a factor to decide whether the horsemen charge home on the objective-form-ation, or swerve away in the face of missile-fire or the hedge of threatening pike-points. Certainly the ability of the infantry should vary according to whether the cavalry are attacking only musketeers, musketeers and pikemen combined, or pikemen on their own. Nor should cavalry be able to do much damage to infantry behind walls or hedges, except with their pistol fire. Defence invariably overtakes and withstands attack in the long run, and in the Thirty Years War infantry soon developed effective defence against cavalry attacks by forming massive 'human forts' of pikemen to hold-off the horsemen, with squares of musketeers alongside, firing in waves resembling the caracole-fire of the cavalry, known in future wars as 'rolling-volleys'. Should musketeers find themselves in danger from the cavalry, they retreated inside the pike square, but the longer range of their muskets usually held-off the cavalry – fire-fights between musket and horse-pistol form an integral part of action in a Pike-and-Shot wargame.

With the exception of the advanced (for their day) Swedish artillery, cannon were unwieldy monsters requiring teams of dozens of oxen to drag them into position on the battlefield before the action commenced, never moving from where they were placed. Unless overrun by the enemy, throughout the battle they belched out their iron balls, cutting bloody swathes through enemy squares. These huge cannon took a long time to load and when wargaming with them only allow them to fire every other move, or even only one move in three. While they are inactive, they can by guarded by formations of musketeers or pikemen and provide yet another combat re-enactment of the tabletop game.

The 17th century matchlock, in wargaming, can be given a range similar to 18th century or Napoleonic muskets, but less frequently to rep-resent their slow rate of fire. Musketeers main-ained a continuous fire by retiring to the rear of the six- or ten-deep file when they had fired, or advancing to the front when they were ready to fire – in the one case the unit gradually retired and in the other it gradually gradually advanced. Or, with the infantry drawn up in small columns six or ten deep with an interval of at least one file between columns, the front rank of each column fired, and filed off right or left through the interval to the rear of the column before starting to reload. This can be reproduced by having the infantry in two or three ranks with intervals between each file (repre-senting six deep or ten deep in real life), and being allowed to load and fire in each move, the front rank only being counted in the volley. Infantry in single rank (representing three deep) are only allowed to load or fire in any one move.

On the wargames table add a bonus-effect to the first volley, because muskets were loaded with care when not under fire whereas all loading afterwards was done to a background of noise, stress and fear. The smart table-top commander will very carefully hoard his first volley.

8
A BATTLE FOR
ALL SEASONS

French 'Battle' Tactics during the Napoleonic period

First: the French tirailleurs engage all along the enemy front supported by medium artillery. Second: French infantry columns, accompanied by light artillery, come forward to attack one or more points in the enemy line. These points may or may not have already received concentrated fire from medium artillery. Third: light artillery go into action at ranges of 150–300yd (135–270m) against the enemy line at the points where the columns will strike. The aimed fire from the tirailleurs, the light artillery close-range fire, plus possible medium artillery fire, and finally the threat of infantry bayonet shock, usually caused the enemy to break before contact.

The American Civil War General William Tecumseh Sherman is reported to have said: 'War is Hell!' although such a commonplace sense of discernment has always been evident even in such lower ranks as Privates who, usually seeing the conflict at much more personal levels than their Generals, can make such an observation with far more justifiable feeling. Of course war is hell, and has been since Pharaoh Ramses II's Egyptians were

The French camp at Auberoche must have looked something like this (Courtesy of Duncan Macfarlane, Wargames Illustrated*)*

A Peninsular War action c1810. (Courtesy Duncan Macfarlane, Wargames Illustrated)

surprisingly assailed by 2,500 Hittite chariots at Kadesh in 1288 BC, the earliest recorded battle in history – and who can say that it is any worse today? A soldier has only one life to give and it is a matter of opinion whether it is preferable to go suddenly by the blinding impact of high explosive, or to be hacked down in medieval close-quarters combat with all the nightmare fear of looking into the blazing eyes and smelling the sweat of the armoured man whose slashing sword is about to extinguish your light of day forever. While killing, and cruelty and mourning have always been constant companions in War, Man's onward March of Civilisation and Progress has armed him with an immense variety of weapons, ranging from the stone axe and bronze sword, through the whole range of gunpowder weapons, to today's nuclear-bomb. What has barely changed, except for being polished with a period gloss, are the tactics with which battles are fought, so that Generals and Officer-Cadets still study and utilise the Battles of History and the tactics of Alexander, Marlborough and Wellington – the sudden unexpected onslaught is no less surprising and shocking today than it was when Cyrus

The French of the Hundred Years' War

The French cavalry of this period were perhaps the finest in Europe in the sense that they were the best born natural handlers of horses and weapons. Unfortunately they were atrociously led and displayed all the vices of the feudal system – lack of control, ill-discipline, impetuosity, disinclination to co-operate, etc.

At Crecy, the French were strung out in a long straggling column without any reconnoitring screen. On 'bumping into' the English, King Philip endeavoured to halt the disorderly mass and concentrate his forces, either for the attack or to have them in hand for an overwhelming assault at day-break on the following day. The mercenary Genoese crossbowmen were the only ones with sufficient discipline to be gathered together and they were sent forward to the assault, only to be decimated by the English

continued

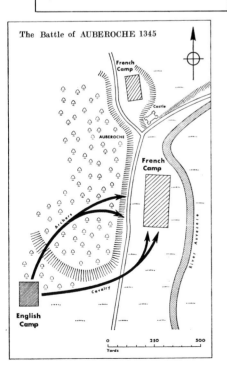

The Battle of AUBEROCHE 1345

of Persia perpetrated the first authentic case of strategical and tactical surprise against Croesus, King of Lydia, at Thymbra in 554BC.

In war, surprise comes in many shapes and sizes, beginning with the large-scale manoeuvres of armies like Napoleon's Italian Campaign of 1796 and his Campaign of Marengo 1800; the advance of Lord Roberts against Cronje in South Africa 1900; the German advance into Belgium 1914 and the triumph at Tannenberg of Hindenburg and Ludendorff; Allenby's Surprise of the Turks at Beersheba 1917; the German Blitzkrieg in France 1940 and D-Day, surely one of the greatest – in every sense – of Military History. Then there is surprise by tactical methods, as at Marathon 490BC; Leuctra 371BC; and methods created by Gustavus Adolphus of Sweden in the 17th Century, and Frederick the Great in the 18th Century; the World War I German attack at Riga in 1817; the first great tank battle at Cambrai in 1917; the final German attack on the Western Front in March 1918; and the German attack through the Ardennes in December 1944. All are stirring examples of the art of warfare and well worth studying; however, they are almost all *big* and few if any wargamers can field armies of model soldiers large enough to realistically reconstruct the actual mechanics of those hallowed operations. But, there *is* a battle absolutely 'tailor-made' for the wargames table, with surprise factors and numbers that cause it to be among the most satisfying wargames of all time. And if the players have actually *walked* the original field – quite unchanged today 543 years after – then it will generate the same reverence felt by the author who, with his 'regular' battlefield-walking comrades, detours miles whenever travelling through Central France to once again chauvinistically revel in this minor English victory of the Hundred Years' War (1335–1453).

It all happened at Auberoche in Gascony on the 21 October 1345, this little-known battle in which a small English force – outnumbered nearly six-to-one – defeated Count de l'Isle's French army by an attack said to be '. . . almost breathtaking in its audacity and dazzling in its brilliance . . .' A French historian of our time has described it as – '. . . un choc terrible . . .' in which the flower of the chivalry of Southern France were killed, wounded or captured, including de l'Isle himself. It is a battle that fulfils all the desired aspects required to hallow its name – a crushing victory achieved by surprise and first-rate fighting deep in enemy territory by a greatly outnumbered army formed of the colourful, almost-legendary English archers, led by a brilliant commander whose tactics and leadership could not be bettered in any age or war. The Earl of Derby, later to be known as Henry of Lancaster, was this Commander, a remarkable man who was continuously engaged in War or Diplomacy from early manhood until his death forty years after the battle. A legend in his lifetime, he fought on the Continent and in the Mediterranean, in crusades and wars in Scotland, Flanders, Brittany, Gascony, and on the sea. When he was forty-six years of age – the same

100

archers and then ridden down by impetuous knights so filled with pride and impatience that they could not wait to fling themselves forward.

Ten years later at Poitiers, the French dismounted their horsemen on the false supposition that this was the only way in which they could beat the similarly dismounted English men-at-arms. But in so doing, they were deprived of their principal asset for offensive action – mobility and shock. The French commander typically gave no thought to any kind of tactical manoeuvre that might turn the outnumbered English from their strong position.

Even seventy years later at Agincourt, the French leaders were still unable to control their impetuous and undisciplined nobility who it is said had 'forgotten nothing and learned nothing'. As a result, the battle went as the others had done when a vastly out-numbered English force of courageous and disciplined archers supporting dismounted men-at-arms in a defensive formation again shattered medieval heavy cavalry and infantry.

In the later stages of the Hundred Years' War the French commanders, at last recognising the superiority of English archery fire-power, avoided attacks against the English in prepared positions and seized all possible opportunities to force them to fight at a disadvantage. Du Gueslin, an outstanding Constable of France, excelled in night attacks and other such stratagems despite English protest that such practices were unchivalrous and 'Unknightly'.

as Wellington and Napoleon at Waterloo, and at the height of his powers – he was entrusted with command of the southern arm of King Edward III's grand offensive against France (remember, this was the year before Crecy in 1346). For this purpose he had but 500 men-at-arms and 2,000 archers, yet in just over three months after landing at Bayonne in early June, had laid siege to and captured fifty fortified towns and castles, won the battle at Auberoche, and laid the groundwork for two future campaigns that were to make Gascony an English province for the next 100 years.

Taking the castle of Auberoche in September, Derby left a small garrison under Sir Frank Halle before marching back to Bordeaux with prisoners and booty. Meanwhile, Count de l'Isle, the French commander defeated earlier at Bergerac, reformed his army and, constructing siege-engines, marched to attack the castle at Auberoche. The chronicler Froissart tells a fantastic story of a messenger sent for help by Sir Frank Halle, being captured by the French, placed alive in a siege-engine and catapulted back into the castle, arriving more dead than alive. But word did reach Derby, who collected a small army of 400 men-at-arms and 800 archers at Libourne, about 65 miles

The field of Auberoche 1345 showing the area of the French camp. Derby's force attacked out of the woods on the right of the road, after laying-up in the woods for hours, watching the French camp

(104km) distant, and mounting them, marched out, ordering the Earl of Pembroke and his force to join him en route. At Bergerac, about halfway, there was still no sign of Pembroke, so Derby carried on marching swiftly under cover of woods until reaching a concealed position in woods less than 2 miles (3km) from Auberoche, without the enemy having any suspicions.

The castle of Auberoche was picturesquely situated on a rocky prominence overlooking the little river Auvezere, about nine miles east of Perigeaux, dominating and blocking the river-valley, narrowed on either side by heavily-wooded slopes. The area has changed but little since 1345, except that the castle is now a ruin and a road has been constructed, between the woods where Derby laid-up and the French camp in the meadows beyond. This fascinating and evocative place in a secluded Gascony valley is not easy to find, reachable by road D.5 from Perigeaux (to Cubiac); at the village of La Change turn left on a minor road (probably marked LAUTERIE) running between two arms of the river, here forming a sort of 'bottle' with Auberoche castle at its neck or narrowest part; the road-sign AUBEROCHE stands at a point about midway on the battlefield, with the French camp on the right and the woods concealing the English on the left. Use Michelin Map No 75 1/200000 Bordeaux-Tulle; find the number '6' on top margin, drop straight down through Savignac; la Bornette; le Bost Vieux; then Lauterie and the battle-field is at the bottleneck of the river-arms.

The French commander de l'Isle with at least 7,000 men, placed his main camp in the meadow to the west of the river (right of the road) where the grassland is about 220yd (200m) wide; with a smaller camp in a still narrower valley on the north side of the castle. Derby's force silently laid-up in the woods about 700yd (630m) from the French camp; horses of the men-at-arms tethered grazing nearby; food for the men had been brought on pack-horses. Throughout the night 20–21 October, the small English force awaited the arrival of Pembroke but dawn came without any sign of him; morning wore on, turned to afternoon still without the reinforcements and Derby, aware he could not wait indefinitely and now without food for his men, decided to attack despite the vast disparity in numbers. He made a personal reconnaissance, groping stealthily through the woods until reaching its edges a few hundred yards from the French camp. He saw a heart-warming sight, French tents and pavilions were spread over the meadow in serried lines, coils of smoke rose from fires cooking the evening meal for the soldiers lounging around their bivouacs, all without armour! He decided a mounted charge from the woods was impracticable as the ground sloped down too steeply, however about 300yd (270m) south of the camp there was a level approach suitable for caval-ry, with a track through the woods leading out into the meadow at this point (it still exists and can be walked). By this track the camp could be charged from the rear, while from the edge of the woods opposite the camp, the archers could give covering-fire until the horsemen reached the tents, then by switching their arrows continuously left, provide further disconcerting covering-fire ahead of their cavalry. In the valley below smoke rose lazily above the tents and English nostrils were tickled by the smell of roasting meat. Cautiously and silently, positions were taken up, and all awaited Derby's signal, when the archers were to unfurl their banners and cry 'Derby! Guyenne!'; then the cavalry would emerge from the wood and charge.

Suddenly and dramatically the still air was rent with ear-splitting cries – 'Derby! Guyenne! Derby! . . .' and archers emerged from the undergrowth, some waved gaily coloured banners – to give the impression of a large force; flights of arrows filled the air – each archer able to put twelve shafts into the air before the first landed. Like an anthill when its stone covering is lifted, confusion reigned in the French camp, all eyes turned to the woods so that none saw the ragged line of cavalry hurtling forward over the 300yd (270km) between the track and the outer tents. Men ran frantically in all directions, trying to don armour as hails of arrows poured upon them – Murimuth, a chronicler of the period, estimated a thousand casualties at this stage. Then the horsemen hit home, surging through the camp hacking and slashing, until the archers had to cease shooting for fear of hitting them. However, the French, in defensive clumps around the few officers who had donned armour or around banners, conveniently formed fresh targets for the relentless searching arrows.

High up in the castle, Sir Frank Halle and his garrison could see the whole incredible scene laid-out before them; while archers added their shafts from the ramparts and men-at-arms hastily buckled-on bits of armour. Brushing aside those French detachments guarding the castle's approach, an avalanche of horsemen erupted out and burst upon the French camp from the other side. It was the last straw, French resistance crumpled and those who could not flee the field surrendered – their comrades in the smaller camp north of the castle, evidently stupefied by events, took no part in the battle and fled for their lives.

In the manner of the Black Prince at Poitiers eleven years later, Derby entertained his captive generals to dinner in the castle that evening. Halfway through, who should appear but the dilatory Earl of Pembroke, to be greeted by a jovial Derby in mock delight – 'You have arrived just in time . . . to help us finish off the venison!'

Even in 'cold' print, this small action stirs the blood and arouses pride in our sturdy ancestors who '. . . like so many Alexanders, have in these parts from Morn 'til Even fought, and sheath'd their swords for lack of argument . . .'

If you want to realise what Military History is all about, pace the field, peer out from those same sheltering woods, let the imagination people the peaceful pastures below with white tents and medieval soldiers sitting around campfires – it is almost possible to scent the aroma of roasting meat – then stride

French Revolutionary Armies
The action was opened by a cloud of skirmishers, on foot and mounted . . . they harried the enemy, escaped from his masses by their speed, and from the effect of his guns by their scattered order. They were reinforced so that their fire should not die out, and they were relieved to give them more efficacy. The mounted artillery rode up at a gallop, firing grape and canister at point-blank range. The line of battle moving in the direction of the impulse given; the infantry and column, for it did not depend on fire, and the cavalry units mingled so as to be disposable everywhere and for everything. When the rain of enemy bullets began to thicken, the columns took to the double-quick with the bayonet, the drums beating the charge and the air reverberating with cries, a thousand times repeated. 'Forward! Forward!'

that same track down which mounted men-at-arms charged, or clamber up to the now-ruined castle on the heights above – it is also what Wargaming is all about!

In many senses it is an 'ageless' conflict, its varied aspects identical whether the small and compact terrain is peopled with archers and men-at-arms; Peninsular War Riflemen, or even Commandos and Paratroopers. It has been fought in this neck of the woods in many guises as befits such a many-splendoured affair and transforms it into A Battle for All Seasons. How about this?

On 2 May 1811, during the Battle of Fuentes de Onoro, Wellington extended his front moderately to the south by moving Houston's 7th Division to Poco Velho and the British cavalry to Nave de Haver, held by Julian Sanchez and his Spanish guerillas. Next day at dawn, under cover of mist, Sanchez was surprised and routed and the cavalry thrown back to Poco Velho, where French infantry in force attacked the 7th Division. Before long the 7th Division and two squadrons of British cavalry which came to their rescue were in a hazardous position, with large forces of French infantry, cavalry and artillery moving against them. Just as the situation looked desperate General 'Black' Bob Craufurd and his Light Division approached Poco Velho, ordered by Wellington to support the 7th and in fine style the Lights relieved the hard-pressed 7th by the simple expedient of substituting themselves in the position, while the 7th retired through them to a new position further back.

But now the Light Division together with Cotton's cavalry were practically isolated in an open plain surrounded by myriads of French cavalry, three infantry divisions and several batteries of artillery. Here was Craufurd's finest hour, in his element having only that morning rejoined his Division from leave; he handled his command with unerring skill, while Cotton's cavalry made partial charges which held back the French guns and Bull's Horse Artillery positioned themselves between Craufurd's squares and fired on all who approached. The whole force retired at a carefully controlled speed and in perfect order, displaying discipline, courage, confidence and physical fitness. This was the immortal occasion when Captain Norman Ramsay's two guns, completely enveloped by French cavalry, cut their way through the encircling mass at a gallop.

Meanwhile the serious fighting in Fuentes had ended by 2pm, the French thrown back with heavy loss, and both armies remained within artillery range of each other without further offensive moves. However, amid the rocks on either side of the Turones River were five companies of British Rifles placed there earlier in the day by Craufurd; having stopped dead swarms of French voltigeurs they were trying to regain British lines, avoiding large numbers of enemy. Four managed it, but a fifth company of the 2/95th Rifles became trapped in the ruins of a castle on a promontory in a bottleneck between two rivers – the Turones and the Dos Casas, at a place called Arbaroz. Hearing of

Major-General Robert Craufurd (1764–1812)

this, 'Black' Bob Craufurd flew into one of those fearsome tempers that gave him his nickname 'Why had one of his companies been left behind . . . was no one going to do anything about it . . . could he have permission to get them out?' he asked the Duke. Unusually mellow after the successful day's operations, the Duke gave permission for a single battalion and a squadron of cavalry to make the effort – and by mid-afternoon General Craufurd had the remaining companies of the 2/95th and a squadron of the 1st Hussars of the King's German Legion ready to march. As they were leaving, Captain Norman Ramsay came up and asked if, as a mark of his respect for the Lights after that day's fighting, could he bring his two guns along? After looking around to see if the Duke had noticed, Black Bob gladly agreed.

To reach the trapped company unseen it would be necessary to take a wide swinging route west of the fortress-town of Almeida, through the foothills and into the thick forests bordering the river, too far for marching men (even at Light Division rate of march) if they were to arrive in time. So, the cavalry 'took-up' the riflemen, each hussar mounting an infantryman behind him on his horse – a method particularly galling as it chafed the soldier's legs and made later marching difficult; and it generally caused him to lose his mess-tins, shaken off by the jolting! At first anyway, those perched on Ramsay's guns and limbers felt they were luckier, until the bouncing bounding guns really got into their stride! Reaching the area of Arbaroz just before nightfall without the French seemingly having the slightest idea of their presence the force laid-up in the woods overlooking French bivouacs that covered the valley floor to north and south of the castle ruins. General Craufurd positioned his riflemen and Ramsay's two guns on the fringe of the wood, placed the German Hussars on a track leading from the woods onto the French rear, and waited for dawn's first light to launch his attack on the hopefully surprised French.

Well, you know what happened next, don't you? History repeated itself without probably a single person on either side having the slightest knowledge they were repeating a legendary action that took place 466 years earlier!

Infantrymen do not come much finer than the Medieval English archer or the famed rifleman of Wellington's Light Division – nevertheless both would have been proud of their descendants who wore red berets and dropped out of the sky in wars far greater and more horrible than anything they could have imagined. So, if the archer and the rifleman could do it at Auberoche and at Arbaroz, then undoubtedly the British paratroopers of World War II – again, without any idea they were repeating history – could emulate their feats at Au-Bar-Roche a day or so after D-Day in 1944. Here is what occurred.

Everyone knows of the 6th Airborne Division's assault on D-Day, on the positions between the Caen Canal and the Dives River; of Pegasus Bridge, Ranville and all those other names now proudly borne as battle-honours on the colours of the Parachute

The Balancing Factor

The average wargamer's enjoyment of a battle often varies inversely as he wins or loses, so that few wargamers will feel inclined to tactically manoeuvre their armies in an historically accurate manner when hindsight tells them that those tactics invariably led to defeat. The French wargamer/commander is going to be very disinclined to throw his columns up the slopes of a ridge when he is aware that the English line, with all their devastating fire-power and subsequent bayonet charge, awaits him below the crest. There is a certain irony in this because it seems to be exactly what occurred throughout history!

But if the wargame is to bear any realistic similarity (other than its title) to the actual historical occurrence then the armies *must* perform in a tactically accurate manner and their wargamer/commander *must* accept that what he gains on the swings he loses on the roundabouts, and that the odds are balanced by providing him with a considerable numerical superiority, for example. Perhaps one of the strongest balancing factors is that of employing a system of 'grading' the commanders of the various armies.

Major-General Richard Nelson (Windy) Gale commanding 6th Airborne Division. Normandy 1944

Opposite:
British Airborne troops in Normandy 1944

Regiment. It was a time of so many tremendous deeds and bravery that events normally sure of a high place in military history were taken for granted. Such an affair was the getting-out of 'A' Company of the 8th Parachute Battalion, trapped in a ruined castle at Au-Bar-Roche near Bures in the area of the Woods of Bavent and the River Dives. All pilots were equal but some were more equal than others, and with the stress of war, darkness and the awareness of the magnitude of what they were taking part in, perhaps there are excuses for those aircrews who overshot the target and dropped their human cargo in the wrong place. So, 'A' Company found themselves in an area quite unlike anything shown on maps and models at briefings, fairly quiet and outside the zone of operations, it seemed to be somewhere well behind Jerry lines. This was discovered when the force gradually gathered together as men filtered-in from all points of the compass, co-ordinating stories of tanks, guns, camps, infantry and other nasties that lay between them and friends. It seemed best to lay-up somewhere quiet, wait for the dust to die down and then get back – the ideal place presented itself when scouts reported a ruined castle on a high rocky elevation overlooking a river. Of course, it was not unoccupied, having been the site of an anti-aircraft battery for the past two years who had to be persuaded that the new tenants would look after the property better than themselves. The resulting fracas attracted the attention of nearby friends of the now-dead German gunners who arrived in such large numbers that the paras found themselves under siege – but a message had gone off before their only radio went 'dis', giving map references and arranging signals in case anyone got around to organising help.

Coming from a similar mould to that which produced the Earl of Derby and General Black Bob Craufurd, Major-General Richard Nelson 'Windy' Gale, who formed and led the 6th Airborne Division in Normandy, was a man whose bluff appearance disguised intellect, tactical shrewdness, knowledge of a soldier's trade – and great courage. He did not like leaving expensively-trained professional soldiers – his lads – out on a limb, and the idea of doing something about it appealed as being a change from the somewhat restrained role his high rank had forced upon him. So, hearing of their predicament, he hastily rustled up a force of paratroopers, men who had been slightly wounded and others separated from their units, plus a couple of gliders carrying armoured Jeeps and took-off at nightfall, to drop and glide-in west of the Bavant Woods and silently disappear into the trees.

Shakespeare said '. . . there is a tide in the affairs of men which, if taken at the flood, leads on to fortune' – and great tactical minds think alike, given similar circumstances and surroundings. So it will come as no surprise to read that Windy Gale, on seeing the German infantry and light armour massing in the valley beneath the prominence on which lay the castle ruins, came up with the sound scheme of covering fire from the wood-

edge while the four armoured jeeps – Browning MGs mounted on bonnet and twin Vickers .303 machine-guns swivel-mounted at rear – burst from the woods onto the German rear, all this at first light, of course. Then a dramatic remorseless charge by red-hatted paras, screaming the North African warcry 'Ahoy Mahomet!' into the flank of the hopefully astonished enemy.

It doesn't have to be spelt out, does it? No need to go into details of how many men in each of the Dakota C.47s (28 actually) nor how there were two jeeps in each Hamilcar glider, towed by Halifaxs. Or how it all worked out, and how the force made their way back to friendly territory, while General Gale was spirited away to his HQ so that neither Monty nor Ike, or any other high-rank knew he had even been away – but his lads did and never forgot!

Wargaming techniques and know-how to fight this stimulating affair will be found elsewhere in this book – in Chapter 5 and details of glider simulated in Chapter 3. The rules governing the other two battles can be those of the specific period currently being used by the wargamer. However, there are obvious circumstances surrounding the basic battle which require consideration if real life resemblances are to be maintained, the primary one being the all-important element of surprise, without which the whole thing falls apart. Because of the omniscience of the wargamer, the factor of surprise is extremely difficult to simulate on the wargames table (it is briefly but helpfully discussed elsewhere in these pages). Two suggestions are offered – first, the 'host-wargamer' takes the attackers and the visiting player the enemy. He is not told the name of the battle, merely that he has a divided force besieging a smaller force holed-up in a castle; the besiegers are arranged in their two camps and, as the battle proceeds, the visiting enemy commander will hopefully be just as surprised (and as indignant) as was the original enemy leader when defeated by such a numerically smaller force. Of course, one can lose friends this way, and it is to be hoped the visitor is no deep student of Military History!

A second method that can be used even with both forces in full view on the tabletop, is for each wargamer initially to draw up a plan of his tactics, down to unit level. Then 8 playing cards, 2 of them aces, are dealt to each commander. Prior to each game-move, he draws a card and, if it is an ace, he is permitted to alter the allotted role of a unit. In circumstances where history indicates one force (or its leader) to be markedly superior, that force is granted a higher proportion of aces than its opponent. Using this method, the complete surprise achieved by Derby at Auberoche can be simulated by his being given 4 aces among his 8 cards, whereas the French are allowed only 1, at most.

I hope you agree it is too good a battle to waste on but a single wargame!

FACILE FORMATIONS – 18th/19th Century Light Infantry

One of the most significant tactical developments of the Napoleonic Era was the extensive emergence of the Light Infantrymen – basically the French skirmisher and the British rifleman. His true role was made possible by the introduction of the rifle, particularly needed by the light infantryman who had to fight as an individual in dispersed order – but because rifles were costly and had a slower rate of fire these specialised weapons were only issued to select units or individuals in line units. However, to give the impression that these highly mobile infantrymen were spawned during the protracted Napoleonic Wars is not entirely accurate, for earlier events had already proved their effectiveness so that major powers had hastened to introduce light troops into their armies. It probably began in 1740, when the Empress Maria Theresa of Austria mustered fierce, colourful semi-brigands, the Croats and Pandours who formed part of Austrian frontier defences against the Turks, to aid in her war against Prussia, France and Bavaria. They made an immediate impact and in 1745 Frederick was forced to detach almost half of his Prussian Army to deal with these Austrian and Hungarian Irregulars when he marched into Bohemia after the battle of Hohenfriedeburg. They became a thorn in Frederick's flesh again at Kolin in 1757 when Croatian Pandours so harassed the solid Prussian columns that they split into three and were defeated in detail – the wild and undisciplined irregulars had little aptitude in formal manoeuvers, but were capable of breaking up regular formations. At this time the Austrians also had Regular Jager battalions – Feldjagerkorps formed of foresters and huntsmen who wore a grey unadorned uniform still in use up to World War One by rifle regiments of the Tirol.

Frederick the Great and Marshal Saxe, the principal leaders of the era, had opposing views on the value of light infantry. Saxe believed in them and one on the earliest units of French light infantry – the Arquebussiers Grassin – played a prominent part at Fontenoy in holding up the Allied right wing. Marshal Saxe had military theories in advance of his time and envisaged mobile legions capable of manoeuvering or quickly concentrating with small and mobile groups of light infantry, taking all possible advantage of the terrain, to harass the enemy and soften them up for the advance of the heavy infantry, when the skirmishers would fall back through the intervals in their formation. Light infantry tactics came naturally to the traditional French style of warfare and good results were obtained by the Chasseurs de Fischer; the Fusiliers de la Morliere; the Volontaires Cantabres, and others.

Frederick the Great, although painfully aware of the ability of light infantry, believed them to be contrary to the Prussian military system, which stressed the crushing of individuality and the training of formations to move with machine-like precision. The independent character of light infantry was totally at variance with Prussian training and could offer frightening opportunities for large-scale desertions.

British light infantry were first raised in the Highlands for service during the '45 Rebellion but did not do very much. These 'specialist' infantry came into their own in North America, first during the Seven Years War and then in 1775 in the War for American Independence. In both these conflicts the close-packed ranks of British and Hessian infantry learned a bitter lesson, their coloured-coats making them a striking target for the rifle-armed American frontiersman who struck with a speed and accuracy while skilfully using all available cover. Becoming acutely aware after Braddock's defeat at the Monongahela in 1755, of the effectiveness of disciplined light infantry and the successful combination of European and Indian fighting as practiced by the Colonists, the British Army was impressed. Irregulars such as Roger's Rangers and the Royal Americans led to the establishment of a light company in each foot battalion, differing in function but not in equipment and discipline, from the rest of the army, and often employed as line infantry. By the time of the American Revolution, it was a customary practice to separate light companies from their battalions and organise them into provisional units.

The traditional conservation of those in authority was shaken and, under the able direction of visionaries like Manningham and Moore, the British Army instituted Light Infantry Regiments. Formed of men trained to act and think as individuals, clothed in dark green and armed with a weapon which out-ranged the conventional musket, these units played an outstanding and significant role in the Peninsular and at Waterloo.

The cream of the British Peninsular army was the Light Division, with their essential role of countering the French Tirailleurs who skirmished ahead of their columns. Thoroughly trained, the British light infantry were able to operate in close order or as individuals in a manner far superior to the average French tirailleur who frequently was only another conscript in a different uniform stuck out in front of

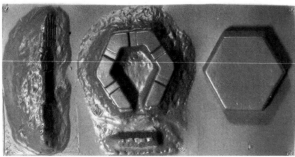

An assortment of commercially-made 'terrain-pieces'

the columns. A scattered rifle regiment could quickly re-form to present a front to deliver the equivalent of musket volley. Impressed by the skirmishing tactics of the Colonists during the American Revolution, Sir John Moore had revolutionised British infantry tactics when he created the Light Division at Shorncliffe. Here Moore applied his theory that line and its fire-power, with the two-rank formation allowing a much wider musketry front with the same number of troops, was the most effective defence against column attack. Realising the need to break up French shock attacks before they could be pushed home, Moore armed his Light Division with rifles, which doubled the distance and accuracy of their fire-power. This weapon, coupled with first-class training in skirmishing, rapid movement and independent action made the soldiers of the Light Division the best infantry in the world.

In 1810, the British Light Division under General Robert Craufurd consisted of:
'A' Brigade – 1st Bn. 43rd; 4 companies 1st Bn. 95th and 1st Bn. Portuguese Cacadores
'B' Brigade – 1/52nd; 4 companies of the 1/95th and the 3rd Cacadores

Two battalions of the 95th, the 5th battalion of the 60th and the Cacadores of the Light Division were all equipped with the Baker rifle, which could be loaded and fired, from a lying position if necessary, in thirty seconds but under battlefield conditions once per minute with an average misfire of one in every thirteen rounds.

It must be appreciated that there was a difference between the Light and Rifle battalions of the British Army, in that the Rifle battalions were armed with the Baker rifle which, although it took longer to load, had an effective range of 200 yards. The Light Infantry battalions had special smooth-bore muskets capable of sustained fire of four rounds in three minutes at a range of 120 yards.

In contrast, the French had light infantry 'thrust upon them' by force of circumstances. The volunteers of the early Revolutionary Wars lacked both the training and discipline to pursue formal drill movements and manoeuvres when under fire. Inevitably, they were repeatedly defeated until, in 1792, it became the practice to push forward the

Prussian Light Infantry skirmishers face French Caribinieres in a Napoleonic wargame (Connoisseur Figures – Holiday Centre-Thomas Collection)

110

most reliable men in a cloud of skirmishers, so forming a screen behind which the nervous battalions huddled in fragile formations.

By 1793 all battalions were capable of acting as light infantry, dissolving into skirmisher swarms as soon as the battle began. These were not special light troops but sections of the Regular infantry who had become more flexible. Later, light infantry battalions, such as the French Légère or British Rifles, usually sent forward no more than half of their strength in skirmish order, with the remainder kept in formed order in the rear to serve as a reserve to support the forward elements, or to form a rallying point if the forward skirmishers were pushed back.

In 1803 there were 26 light regiments in the French army but by 1813, in spite of no less than 127 new regiments having been formed, the proportion of light units had diminished to about a sixth of the total infantry strength. During the three years that followed 1805, line battalions were reorganised and given six companies – 1 grenadier, 4 fusilier and 1 voltigeur (light infantry). The 1808 establishment gave the light infantry battalion 1 carabinier company (equivalent to line grenadiers); 4 chasseur companies (fusiliers) and 1 of voltigeurs or skirmishers.

The counter-actions of tirailleurs and riflemen can be seen in one of Wellington's early Peninsular battles in August 1808, when he had positioned his 16,500 on two long hills at Vimiero, where they were attacked by 13,000 French in two columns shielded by tirailleurs, supported by mobile field artillery and with cavalry protecting their flanks. The tirailleurs were so troubled by the riflemen that they were unable to adequately protect the columns and the French gunners were harassed so that their fire was light and ineffective.

One Light Battalion to four or five Line Infantry Battalions is a reasonable proportion in a wargame army. Light Infantry can be used as a screen for infantry attacks but are most valuable on defence as they can move and fire, firing then retreating, etc. They can hover around infantry columns, harassing them with fire so that the infantry are finally forced to deploy to deal with them. Their open order skirmishing formation makes them very valuable in wooded or close country where a column or line of infantry are swiftly disordered. During the Napoleonic period, skirmish lines had one or two men per 5 yards of front; firing lines four to eight men per 5 yards of front and battle lines (two or more lines) sixteen or more men per 5 yards of front.

When wargaming, Light Infantry on the march will form up as a body like any other infantry unit and march at the same rate, but when they operate individually as skirmishers they should be given a greater move-distance. This is counter-balanced by a loss of solidarity and cohesion because they must operate at least an inch apart, whereas Line Infantry of this period fought 'base-to-base'. Light Infantry operating less than an inch from their fellows will be treated as ordinary infantry and will move at the same speed. In the Peninsular the British marching pace was 120 per minute which was matched by the French so that battlefield mobility was roughly equal, therefore move-distances will be the same.

One of the duties of the Light Infantry skirmisher was to pick off individual officers, and as most rules have a penalty for loss of officers, this can be a vital factor. A Light Infantryman can specify the officer at which he is aiming but it should not be certain that he actually hits him, a dice is thrown and if it shows up 1 or 2 then it is the man immediately on the officer's right, 3 or 4 it is the officer himself, 5 or 6 it is the man on the officer's left who is hit.

In mêlées their properties are those of ordinary infantry although, when caught in the open by a formation of infantry or worse still by cavalry, their extended order makes them extremely vulnerable and they are very lucky not to be cut down. Rules should be slanted to cover these factors.

Light Infantry in wargames introduces a completely new dimension, allowing considerable expression of temperament and personality – both of the wargamer and his obedient miniature units!

95th Foot Riflemen skirmish in the Peninsular 1810 (Hinchcliffe Figures)

9
THEY CAME IN FROM THE SEA
Commando Raids as Wargames

The Commandos came into being literally overnight when the fertile fancy of Winston Churchill was taken by the suggestion of morale-boosting raids on German occupied Europe. In June 1940, it was decided to form 10 Commandos each of 10 troops and volunteers were called for from military commands in the UK. The men were to be fully-trained soldiers, physically fit, able to swim, and immune from sea and air sickness – among the qualities demanded of them were '. . . courage, physical endurance, initiative and resource, activity, marksmanship, self-reliance, and an aggressive spirit towards the war' besides having the ability to stalk, to move across country silently and unseen by day or night, and to be able to 'live off the country' for considerable periods. The response was immediate and overwhelming and very quickly highly suitable commanding officers had been appointed and were carefully selecting their officers and men from the numerous volunteers.

Fearing confusion with the infamous Nazi SS, their original title of Special Service Battalions was speedily changed to Commando, because their role was to closely resemble that of the Boer mounted troops in the South African War. It was an inspirational choice that grafted a new accolade onto an already auspicious title. Called by Winston Churchill 'A steel hand from the sea', they gave heart to their beleaguered countrymen by daring and courageous raids on enemy coastlines. They landed and fought on French soil; north of the Arctic Circle; slipped ashore on Axis-controlled Atlantic, Mediterranean and Pacific shores; spearheaded amphibious assaults on D-Day, in North Africa, Sicily and Italy; and were employed as 'elite' assault troops in North-West Europe. The Rangers, their American counterparts, similarly fought in North Africa, Sicily, Italy, Normandy and the Pacific.

Initially they fought without artillery and heavy weapons, in unsupported surprise actions, relying on their well-trained abilities with personal arms. Needs changing, they fought relatively 'conventional' actions in far larger formations; usually successfully and always displaying supreme daring and bravery.

Their raids and small-scale operations admirably lend themselves to reproduction on the wargames table, where the numbers of men involved can realistically be reproduced in miniature. The comparatively confined battle-areas enable wargames to be fought on a 'man-to-man' basis, ie, 1 man in

At Vaagso 1941, Commandos turn a captured German field-piece against its former owners

Commandos on the Vaagso Raid, 1941

113

Using 1:300 scale models, a direct relationship to scale is only possible between small numbers of AFVs. For a general action, a ground-scale of 1:2000 is suggested enabling tanks to be manoeuvred out of effective range of armour-piercing ammunition. This scales out at 1 mile per hour; equivalent to ¼in (6mm) per move. Due to the limitations of scaling-down at 1:72 scale, it is unrealistic to allow more than one section or troop of tanks on either side and at 1:300 scale a battalion is overcrowding the battlefield!

The time period covered by a move representing 30 seconds of real life, is important, for example, rates of fire vary considerably between different weapons – mortars are capable of 12 to 15 rounds per minute whilst a heavy gun of say 100mm would be lucky to get off 3 or 4 shots in the same time. Field artillery, unless saturating an area, has to wait until the forward observer has reported the point of impact before fitting its second round. All these influence the time period each move represents and thirty seconds seems to allow optimum rates of fire and is recommended when firing tank guns. One man could quickly and easily load calibres up to 57mm, with 75–90mm shells being about the biggest he could handle. The 122mm gun in the Russians JS3 had to be loaded with two-part ammunition, just as does the gun in the British Chieftain tank.

Inevitable distortion of scale on the wargames table interferes with ground-distance realism so that, in 1:76 scale, a 1,000yd (900m) wide river such as the Lower Rhine scales down to *continued*

real life = 1 model soldier on the tabletop battlefields. The size and topographical features of their historic battlefields are capable of being reproduced on the wargames table; almost all Commando fields seem suitable in area and frontage for this purpose. Every one of these actions possesses tactical and human interests making it worthy of simulation, besides bringing to the wargamer a new and fascinating facet of warfare, involving soldiers, sailors, marines, vehicles and boats of all types; and weapons which can be authentically simulated within the confines of an orthodox wargames table. Here the wargamer is not restricted to relatively conventional military formations that dictate their own movements and tactics, but is handling forces of widely dissimilar numbers and types. As on D-Day, these highly trained elite troops were sometimes employed in operations which, although requiring high-grade infantry, were essentially conventional and different to their usual surprise 'one-off' operations.

Acknowledged the greatest Commando Raid of all, the attack on St Nazaire on 27–28 March 1942 can be reconstructed as a most stimulating tabletop wargame. Attempting to simulate this historical conflict on the wargames table, remember that, like the majority of Commando operations, it includes numerical disparities, surprise factors, varying qualities of troops, inequalities of weapons and morale effects difficult to handle under normal wargames rules. St Nazaire, in particular, reveals gross disparities in respective numerical strengths, neutralised to a large extent by the essential element of surprise, the numerically fewer raiders selecting their own time and place. There are occasions in battle when a commander takes a deliberate calculated risk which no normal set of wargame rules would allow any chance of success; the story of Commando warfare demonstrates many such gambles that came off and rules should include purposeful biases for reasonable simulation on the

Wargaming a Commando raid on the French coast

wargames table. Most battles, particularly the dashing, surprise-ridden Commando operation like St Nazaire, require 'local rules' that make allowances for factors peculiar to a specific operation – none but the most wildly 'cooked' sections on morale will give much chance to the small groups of Commandos with their pistols at St Nazaire, taking on multiple quickfiring guns. There are many 'out-of-the-normal-run' incidents which make hay of conventional wargames rules, formulated on a norm of warfare where extraordinary feats of bravery or cowardice can occur but only when governed by incredible dice throwing! To cater for these exceptional incidents, which abound in Commando actions, 'local' rules are about the only suggested solution.

On the other hand, in the interests of an accurate simulation, unusual situations could be recognised and acknowledged by all participants, who resolve them in a reasonable and equitable fashion whether or not the prevailing rules allow for them, because no set of rules can cover *all* battlefield eventualities. Commando operations involve troops of varying qualities, highly-trained, superbly moraled 'elite' troops, often pitted against run-of-the-mill soldiers of the Line, perhaps conscripts of low medical category serving in what is considered a rear area.

Every factor and facet must be practically considered if the reconstruction of such operations is to be more than a matter of giving a name to a wargame.

Wargaming a Commando raid on a harbour area

about 36in (91cm) in width. Villages and groups of buildings are affected in ground-area so that one house might represent a whole hamlet of 30 to 40 houses. This produces another problem – the essence of good tank tactics is, by taking advantage of every undulation in the ground to obtain 'hull-down' positions – the tank taking up a position with its hull concealed from the enemy and only its turret exposed for firing purposes. If realism is to be maintained, the scaled height of undulations must be kept constant and not distorted in the same ratio as the horizontal scale, which means that reproductions of actual battle terrains cannot be reasonably scaled, having to be represented by a tabletop terrain that is similar in appearance.

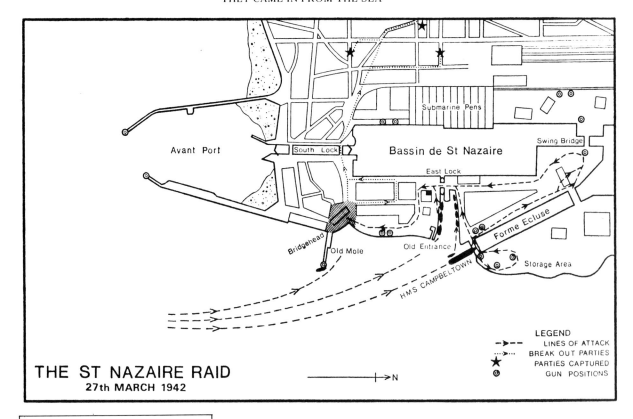

THE ST NAZAIRE RAID
27th MARCH 1942

LEGEND
LINES OF ATTACK
BREAK OUT PARTIES
PARTIES CAPTURED
GUN POSITIONS

Keeping in Touch

There are occasions during wargames battles and campaigns when, as would occur in real life, the commander of a force needs to communicate with another distant group of his own side. Today, and in recent wars, this is done by radio – and sometimes that is not as easy as it sounds – but in past times commanders had to write their message and despatch it by mounted courier. This didn't always work out either – the Light Brigade disastrously charged at Balaclava in 1854 because their commander Lord Lucan could not understand the message sent down from the Heights by Lord Raglan and borne by Captain Nolan. On the wargames table the eventualities of sending messages are simulated by the use of a set of Courier Cards, each bearing a different instruction or condition *continued*

The Raid on St Nazaire – 27–28 March 1942

During World War II a British commando and Naval raiding party attempted to block the German-held port of St Nazaire, which had the only dock outside Germany large enough to accommodate the two great German battleships Bismarck and Tirpitz; the intention was to prevent the Germans forming an Atlantic raiding force based on St Nazaire and Brest. Because of its importance, St Nazaire was well defended with coastal and dual-purpose anti-aircraft guns, plus a garrison of 3–4,000. The 6 miles (10km) of river approach were negotiated by the attacking vessels moving, at high tide, over the sandbanks in the middle of the estuary. The air raid laid on for 40 minutes before the actual assault, to divert attention, was not pressed home because of low cloud, but it alerted the defenders.

The attacking force consisted of 280 commandos and 350 naval personnel carried in the destroyer *Campbeltown*, 16 motor launches, 1 motor torpedo boat and a motor gunboat.

In the early hours of the morning of 28 March the attacking force made its way unobserved up the estuary to within less than 2 miles (3.2km) of the objective when it was illuminated by searchlights. False radio signals sent out in German won a few minutes' delay, enabling the *Campbeltown* to pass the last point at which she could be fired on by the main coast batteries, so that she was exposed only to the fire of light weapons for the last 5–6 minutes of the run in. When the enemy's guns opened fire, all the guns on the attacking vessels replied. The *Campbeltown* moved at top speed to break through the torpedo net guarding the lock gate (of the Forme Ecluse) and crash into it with her bows stuck fast as she was scuttled and sank.

The assaulting groups were divided into Headquarters, Assault, Protection and Demolition parties, with a Special Task party and a reserve of 12 men. The job of assault parties of 2 officers and 12 other ranks each, armed with Tommyguns, Brens and rifles, was to form a bridgehead and perimeter, blocking all lines of approach from the main town; to clear the enemy from the outer harbour and destroy the guns there; eliminate the gun positions on each side of the main drydock entrance; to put the guns on the roof of the pumping station out of action; and to destroy the flak towers at the north end of the dock. The Special Task party was to destroy 2 guns between the Old Mole and the Old Entrance and to damage any ships they came across. The Protection parties, armed like the Assault parties, and consisting of 1 officer and 4 men each, were to guard the Demolition parties, whose strength varied according to their task (their only personal arms were revolvers). Their role was to place charges below water level at the main gate to ensure its destruction after it had been rammed by the *Campbeltown*, to blow up the pumping station, to destroy both winding houses and to smash the inner drydock gate – if successful this would put the great drydock out of action for months. Then with only 1½ hours for their task, if time and circumstances permitted, they were to blow up the bridge connecting the dock area with the mainland and, when the troops had withdrawn to the Old Mole (the place of re-embarkation), to blow the bridge and dock gates connecting the Old Entrance with the Bassin St Nazaire to prevent any counterattack from making contact and also to make the entrance impossible for U-boats. Finally, the two bridges and sets of dock gates and the lock connecting the outer harbour with the Bassin St Nazaire were to be destroyed.

All but 1 of the 7 motor launches whose job was to land troops on the Old Mole were destroyed or disabled before they could do so. Of the 6 MLs that were to land their troops at the Old Entrance, 1 did so, and 2 others missed the entrance but regained their bearings, and turned to land their commandos. The remaining 3 MLs in this column were hit before landing their troops.

As soon as the *Campbeltown* struck, Colonel Newman, the Commanding Officer, and his party went ashore at the Old Entrance from motor gunboat 314, and made straight for the point selected as Headquarters, where they were to meet another party – which did not arrive. Soon the small group was under heavy fire, but was relieved by the arrival of a small group with a 2in mortar, which temporarily silenced the guns on the submarine pens.

That part of the force put ashore at the Old Mole encountered heavy and stubborn opposition from two defended pillboxes

controlling the speed at which the message travels to reach its destination. Ruling that couriers ride at 5 miles (8km) per Game-Move, workout how far apart are the two groups to reach an estimated travel-time, then draw a card and act accordingly. The example-cards given here are based on two forces 10 miles (16km) apart.

Card No. 1 COURIER'S HORSE LAMED, TAKES THREE MOVES TO REACH MAIN BODY

Card No. 2 COURIER'S HORSE VERY FLEET AND DOES JOURNEY IN ONE MOVE

Card No. 3 COURIER ATTACKED BY BRIGANDS AND MAKES DETOUR THUS HE TAKES THREE MOVES TO REACH MAIN BODY

Card No. 4 COURIER MEETS PATROL FROM MAIN BODY AND MESSAGE REACHED THEM IN ONE MOVE ONLY

Card No. 5 COURIER MEETS CAVALRY FORCE WHO CARRY ON TO SCENE OF ACTION ARRIVING AT END OF 2nd MOVE. COURIER ARRIVES MAIN BODY AT END 2nd MOVE

Card No. 6 COURIER'S JOURNEY UNEVENTFUL – REACHES MAIN BODY IN TWO MOVES

The Old Mole jutting out to sea at St Nazaire down which the Commandos assaulted. The old entrance is at bottom left of picture

St Nazaire 1988. On the centre left is the Forme Ecluse, the dry dock, and the bridge across which people are walking is the dock gates through which the destroyer Campbeltown *smashed, approaching in the same direction as the boat coming in from the right*

(which were never completely put out of action), from guns in a high building near the submarine pens, and from machine-guns mounted on the roofs of buildings. Heavy explosions indicated that the demolition parties were carrying out their tasks, although opposition was heavy and considerable fighting took place.

The commando group that went ashore from the *Campbeltown*, organised in a HQ and 4 parties, had a number of men wounded before getting ashore, but completed all its tasks successfully within the allotted time. Then it withdrew as planned to the bridge over the Old Entrance where, in due course, it was joined by the survivors of the parties which had got ashore at the Old Entrance and the Old Mole.

It was obviously impossible to disembark from the Old Mole, so Colonel Newman and about 50 men, many slightly wounded, after grouping at Headquarters, split up and tried to make their way through the town into the open country, intending to return to England through France and Spain. In the event the majority of them, including Newman (who was awarded the VC), were captured.

HMS *Campbeltown* blew up at 10.30am, shattering the dock gate and so damaging the dock that it was never repaired by the Germans.

The British casualties were 144 killed or missing and 215 taken prisoner. The German casualties are believed to have been about 70 killed and an unknown number of wounded.

The battle can begin by involving only those troops who disembarked accepting that those on vessels known to have been sunk took no part in the simulation. Another way is for the raiding force to be marked on the map at a specified point in the river, and then proceed, with vessels taking casualties from fire as they near the objective. Military Possibilities and Chance-Cards can recreate the historical situation, in which the German coastal batteries opened up very late, so allowing the force to get inshore.

The targets of the original raid are listed, but the inclusion of all of them in the tabletop reconstruction will be impossible. The targets chosen will be named on the commandos' map and the prime objective will be to destroy them within a specified time. The fate of the raiding party is a secondary consideration, nor is there any value in killing large numbers of Germans.

The 3–4,000 German soldiers were obviously not all in the immediate area of the docks at the onset of the raid; though sentries must have been on duty, pillboxes and other defence posts manned, and coastal batteries and flak towers obviously alerted by the earlier air raid, or from the moment firing first began in the river. Therefore, these posts can be manned by adequate numbers of Germans, while the remainder of the force is considered to be 'in barracks' somewhere in the town, so that, once alarmed, it will have to be assembled and transported to the area of fighting. On a Time Chart mark (a) the time at which the alarm reached them, (b) delay while they are rising, dressing, arming themselves and forming up, and (c) the time taken to embus or proceed on foot to the dock area, considered to be the German 'baseline'. Up to that point the Germans will be moving 'on paper', and they will only manoeuvre tactically, and as figures on the table, when coming forward from their base-line.

This aspect of the battle can be simulated by having a German 'controller' in another room, with an accurate scale map of the area, a Time Chart and a scaled table of movement. Fed with notes sent back from the points of conflict, he will act by sending troops wherever required.

Rating of Commanders and Observations

Colonel Newman and Commander Ryder, the Naval Commander, will be 'above average' and the German commander, who apparently did nothing particularly wrong or inadequate, can be considered as 'average'.

The morale and fighting qualities of the raiders must always be higher than those of the defending Germans for the following reasons:

1. The attackers were selected, specially trained assault troops.
2. The defenders were run-of-the-mill garrison troops, who might even have been low category men invalided from the Russian Front.

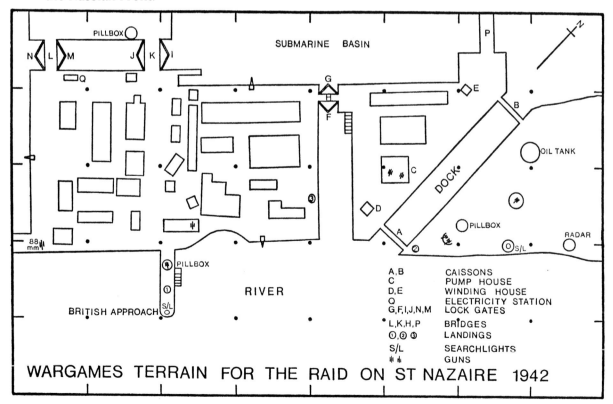

A,B	CAISSONS
C	PUMP HOUSE
D,E	WINDING HOUSE
Q	ELECTRICITY STATION
G,F,I,J,N,M	LOCK GATES
L,K,H,P	BRIDGES
①,②,③	LANDINGS
S/L	SEARCHLIGHTS
	GUNS

WARGAMES TERRAIN FOR THE RAID ON ST NAZAIRE 1942

3. The attackers had darkness and surprise as their very powerful allies.
4. The defenders were detrimentally affected by the darkness and surprise, which would make them apprehensive and lower their morale and fighting qualities.

These factors should have speeded up the raiders' reactions, so that, in any face-to-face confrontation, they are always assumed to have fired or taken offensive action before the Germans. The reconstruction will end if the raiding party can get a substantial proportion of its party, with wounded etc, taken off from the Old Mole; or at a point where all its possible targets have been eliminated; or when it is so decimated that few if any of the targets can be destroyed.

There is scope in this operation for some ingenious rules concerning the simulation of explosives and their effectiveness; there should be specified time-factors allowing men to be clear of a building before the charge explodes.

Construction of Terrain

It is not one that can be knocked up in an odd hour before the game starts, because it requires buildings, docks, etc. The dock area can be chalked out on a wargames table or represented by a sheet of card, hardboard or plywood, with the outlines of the docks and jetties, etc, either cut out or marked in. The buildings can be made from card or blocks of wood; or there are excellent plastic kits of factories and buildings used in model railway construction. The destroyer *Campbeltown*, or destroyers like her, exist as kits; the other vessels can be scratch-built, made up from kits or card, or hardboard shapes can be used. The soldiers themselves are plastic HO scale, cheaply and readily available from sets put out by Airfix and ESCI.

The Spanish Civil War

The Spanish Civil War, which broke out in 1936, was fought with a queer blend of relatively ancient and modern weapons, as, desperate for any armaments, both sides, particularly the Republicans, used old weapons of World War I, including even the earliest armoured fighting vehicles, such as French Schneiders and Renault FT17s, with some converted armoured cars. This makes the Spanish Civil War a unique opportunity for a free-and-easy sort of wargaming, using almost any troops and equipment from the middle of the World War I period to the early days of World War II.

Other countries intervened in the struggle in a most exceptional manner, using Spain as a testing ground for the equipment likely to be used in future wars. To test out their armoured vehicles under battle conditions, the Italians sent Franco a battalion of Fiat CV33 tankettes, with a few CV33 flame-throwers. They were joined by the Condor Legion, equipped initially with Renault M17s or FTs, but later with Panzer Is. This German formation contained four battalions, each of three companies of 50 tanks. The Russians shipped large quantities of arms to the Republicans, and in 1936 they provided 65 T26Bs and 15 BA10 armoured cars. So successful were the T26s against Panzer Is and CV33s, being invulnerable to their machine-guns, that the Condor Legion offered rewards for the capture of T26 tanks for their own use. In 1937 the Russians shipped in BT5 tanks to the Republicans. The biggest tank attack of the war was in March 1937, when the Italians sent in 250 CV33 tankettes, supported by infantry and artillery.

10
HOW ABOUT RE-FIGHTING THE FIRST BATTLE OF ALL TIME?

The first great Egyptian General was the Pharaoh Thothmes III who ruled for 50 years from about 1500BC. He greatly expanded the small Egyptian standing army, grouped infantry into units with standards each in the form of some sacred object or device borne upon the points of a spear by an officer chosen for his bravery. Mercenaries were employed and units using different weapons were trained to co-operate with each other. There were units of archers and slingers, heavy infantry and spearmen who carried large shields that covered them from neck to ankle. Little is known of the use of cavalry by the Egyptians. With these forces, Thothmes III fought the battle of Megiddo, better known as the Biblical Armageddon. He certainly had his army under sufficient control to permit tactics that gave the Egyptians overwhelming victory. At this time the Egyptian troops appear to be well disciplined and drilled, and were the first fighting force trained to respond to the sound of trumpet or drum.

It was under another Pharaoh, Rameses II, that the army of Ancient Egypt reached its peak. Realising that chariots changed the tactics of war by making battles mobile and fast moving, the Egyptians equipped their forces with war chariots made entirely of wood and leather; light vehicles with spoked wheels capable of manoeuvring at high speed and so light that a man could carry one on his shoulders. Chariots played a great part in Egyptian warfare of this period; used in large numbers, they relied on shock effect as well as fire-power from the archer carried in each vehicle. It was a time when, against any but highly disciplined infantry, the massed chariot attacks were most effective, only being checked by counter-charges which resulted in a melee of rushing vehicles flashing past each other in a cloud of dust, arrows and javelins.

Military uniforms were simple in style, consisting of a striped bonnet head-dress and a waist girdle of strips of leather sewn to one another. Iron or bronzed helmets and leather jackets covered with metal scales were reserved for higher ranks and the mercenaries. The ordinary soldiers carried an ox-hide shield stretched on a light wooden frame with a metal disc about 8in (20cm) across in the middle and endeavoured to take the impact of arrows and spears upon this metal disc. In addition to the khopshou, a hatchet with a broad curved blade, the soldiers were variously equipped with axes, bows, daggers, short swords,

The Roman Legion's Method of Attack
A legion consisted of ten cohorts each of 600 men, divided into ten 60-man groups set out three-deep on a frontage of 2,000 ft (600m). In the attack, the front line of cohorts advanced until, at 200 ft (60m), the first two rows of each cohort ran forward and flung their pila when within 50 ft (15m) of the enemy. Next, the rear rank did the same, then the front ranks were among the enemy thrusting with their swords. Meanwhile, the second line of cohorts was preparing to move forward in like manner, while the third line of cohorts waited in reserve, either to be used for 'mopping-up' after victory, or covering the retreat if their comrades were defeated. Throughout, the slingers maintained a hail of stones on the enemy's rear and the auxiliary cavalry drove in on the enemy flanks to bunch and disorder them for the frontal attack.

javelins and the 6ft (2m) long short spear. On the battlefield, light infantry armed with bows, slings and daggers went forward in advance endeavouring to disorder the enemy with a hail of missiles. Next, the heavy infantry, their spears held out in front of them, drove like a wedge into the enemy's centre. At the same time, massed chariots on either flank bore down at full gallop in a charge so perfectly timed and drilled that not a single chariot would break the line.

Style of Fighting

In battle, spearmen stayed together in large groups, lightly armed archers and slingers swarmed out in front whilst the nobles, mounted or in chariots, surged around on the flanks. Manoeuvre was usually accidental as the armies advanced towards each other, both keeping up a harassing fire until the chariots and cavalry charged. Then the light troops drifted to flank and rear through ill-defined gaps between the large groups of infantry. Sometimes the chariots and the horsemen so frightened the enemy that they turned and ran so that the battle turned into a murderous chase. Otherwise, both masses clashed together in a huge melee with the line swaying backward and forward until one side suddenly broke – then the chase began.

In 1292BC Rameses II, an energetic young man of great ability, became Pharaoh of Egypt and set about restoring the Egyptian Empire to its former glories. His main enemies were the Hittites, who were strongly based in Syria, with Kadesh as the main bastion of their southern frontier. In the spring of 1288BC Rameses took the field, and by mid-May was traversing the valley of the upper Orontes, a day's march from Kadesh in the plain below. In order of march the Egyptian army, 20,000 strong, was formed in 4 divisions:

The Division of Amon, with the Grand Guard led by the Pharaoh
The Division of Re
The Division of Ptah, including the Base Troops (recruits and young men)
The Division of Sutekh

When they reached the ford near Shabtuna, the troops halted to water and feed, and became strung out, causing a gap of from 1 to 2 miles (1.6 to 3.2km) between each division. 3 miles (5km) from the ford the Forest of Baui slowed the column and allowed the division of Amon to get even farther ahead as the impatient Rameses, anxious to besiege Kadesh, marched on without waiting for the remainder of his army. By mid-afternoon, after covering 15 miles (24km), Rameses reached a suitable camping area 1 mile (1.6km) west of Kadesh and some 8 miles (13km) north of the ford. Camp was pitched, with a surrounding zariba of shields and Pharaoh's pavilion in the centre, chariots were parked and horses picketed.

Opposite:
Ancient Egyptians (Miniature
Figurines)

continued

**The Roman Civil War
50–44BC**
The death of Crassus at Carrhae in 53BC brought Julius Caesar and Pompey into conflict for Rome; ordered by the Senate to disband his army and return to Rome Caesar massed 9 legions of 40,000 veterans, 20,000 auxiliaries and cavalry. Pompey had 2 legions in Italy, 7 in Spain and controlled perhaps 10 or more additional legions in Asia, Africa and Greece, plus many auxiliary troops; and the Roman Navy. Caesar marched rapidly southward along the Adriatic coast, collecting recruits and reinforcements causing Pompey to abandon Rome and flee to Greece. Before marching overland to reach Pompey, Caesar had to protect his communications by defeating Pompeian forces in Spain. Leaving Mark Antony in Italy and sending Gaius Curio to Sicily and Africa, Caesar's army went ahead to seize the passes over the Pyrenees; Caesar left Gaius Trebonius to besiege Massilia (Marseilles) with 3 legions and hastened off to Spain. A Pompeian army of 65,000 men under Afranius and Petreius waited for Caesar at Ilerda (Lerida) and in the campaign that followed, Caesar with 37,000 men manoeuvred to avoid Roman fighting Roman until causing the enemy to surrender, many of their men joining his own army. In Africa Curio was defeated by a Pompeian force under Attius Varius in alliance with Juba, King of Numidia. After Brutus had won a naval victory over the Pompeian fleet, Caesar secured the surrender of Massilia; then heard that his small navy in the Adriatic had been defeated. Despite Pompey's control of the sea, Caesar transported 7 legions and some cavalry (about 25,000

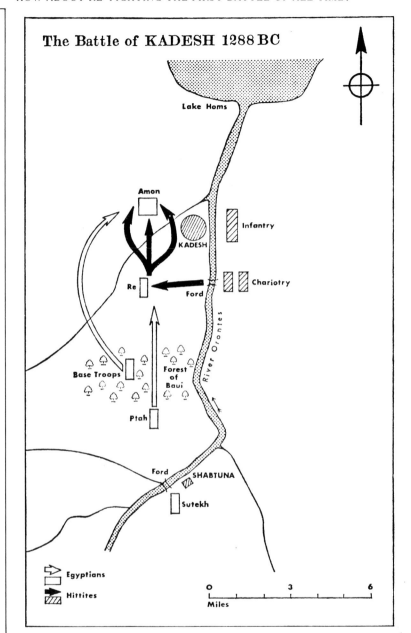

Suddenly a rider galloped into the camp crying that a host of Hittite chariots had crossed the river to strike the divided Egyptian army, hitting the unguarded flank of the Division of Re and routing it. Mutallu, the Hittite commander, had completely outwitted Rameses with false information of his whereabouts, inducing the Egyptians to move along the left bank of the river while, on the right bank, the Hittite army kept the hilltop town of Kadesh between them and the enemy. In this way, completely unseen, they were able to close in from the south-east before fording the river 2 miles (3.2km) south of Kadesh to cut the Egyptian army in two and isolate Rameses, who was now opposed by vastly superior numbers.

He had only sufficient time to send his Vizier for help before fugitives from the Division of Re came pouring into the camp carrying men and tents before them. Then, as their pursuers came on the disordered scene, Rameses rallied his men and cut his way out to the north. Instead of pursuing, the Hittite charioteers began to plunder, as did the infantry formations following them up. Given vital time to regroup, Rameses flung his chariots in charge and countercharge on the weak enemy eastern flank by the river.

Some 4½ miles (7km) farther south the Vizier's chariot, keeping on the fringe of the battle, had come upon the detachment of Base Troops just emerging from the Forest of Baui; they were ordered to strike off to the left and, from the west, attack the Hittites in the captured camp. Continuing, the Vizier next met the Division of Ptah, which he led in a frontal attack from the south. Still south of the Orontes, the Division of Sutekh was too far away to intervene in the battle.

Arriving at the camp, the Vizier led forward the Na'Arun troops, a crack Canaanite unit, in phalanxes 10 ranks deep, to fall upon the now disorganised Hittite troops. Heavily engaged with Rameses to the north, assailed by the Base Troops from the west, and now attacked by these crack troops from the south, the Hittites looked around for support; but none was forthcoming, as the remaining 6,000 Hittite infantry stayed in its ranks on the far side of the river. Mutallu, having already committed all his chariots in the first attack on the centre, probably realised that his infantry were powerless against the Egyptian chariots. It is recorded that the Hittite King was drowned attempting to cross the river – Egyptian reliefs show him being pulled feet first out of the water.

The Hittite charioteers wavered, then turned and fled in a terrible rush for the ford, which was quickly blocked by a mass of struggling men and horses, and overturned chariots, into which the Egyptians poured a pitiless stream of arrows until there was not a single Hittite left on the west bank. The Hittite survivors retired into Kadesh to prepare for a siege, but the battered Egyptians also pulled back without attempting to take the city – so the battle ended indecisively.

Reconstructing the Battle

Although this battle was strung out over a front of about 2 miles (3km), it has to be constricted to fit into a single wargames table, with the Hittite flank attack on the Division of Re taking place at one end and with the Amon camp at the other, allowing space for Rameses' counterattack and the final fighting when the Base Troops and the Divisions of Ptah arrive.

The organiser of the wargame should take the role of Mutallu, the Hittite leader, thus preserving surprise as the Egyptians lay out their forces with the Division of Amon in camp and the Division of Re approaching (see terrain map). Ideally, separate wargamers should command the Divisions of Amon and Re.

men) in every available vessel to Greece, landing at Dyrrachium (Durazzo). Pompey's army of 100,000 although many veterans, were inferior to Caesar's army. Antony arrived with his army and after skilful manoeuvering, Caesar invested Pompey at Dyrrachium but was defeated. He regrouped in Thessaly, with 30,000 infantry in 12 thin legions plus 1,000 cavalry against Pompey's 60,000 infantry and 70,000 cavalry and at Pharsalus on August 9th 48BC Caesar employed brilliant tactics to defeat Pompey who fled to Egypt, where he was assassinated. His supporters in alliance with the Egyptians carried on the fight; there were Naval battles at Alexandria where Caesar lost his ships before slipping out of the city to join Mithridates of Pergamun who had marched his small army from Asia Minor. In the Battle of the Nile the allies completely defeated the Pompeian/Egyptian armies. The Pontic Campaign of 47BC in Asia Minor ended with Caesar victorious at Zela. After quelling a mutiny of his veterans in Rome, Caesar sailed for Africa to defeat a 50,000 strong Pompeian Army with a Numidian force of equal size, supported by a formidable fleet at Thrapsus. Then Caesar sailed to Spain with 40,000 men fought the young Gnaeus Pompey and the renegade Labienus with 60,000 men, in perhaps the most bitterly contested of all his battles at Munda on March 17th 45BC. This concluded the Civil War.

continued

Boadicea's Revolt AD61

In AD61 the frontiers of the Roman Province of Britain had advanced to the line of a road connecting Gloucester and Wroxeter, with highways connecting important towns, over which the Romans maintained their communications by hard-riding horsemen. The Province of Britain was governed by Suetonius Paulinus with forces disposed at Wroxeter, the most powerful fortress with a rampart 2 miles (3.6km) in diameter, and the largest settlement in Britain dominating the upper Severn, housing the XIV Gemina and XX Valeria legions together with four auxiliary cavalry regiments and six auxiliary cohorts of infantry (17,000 men). At Gloucester were II Legion Augusta plus 3,000 auxiliaries formed in three cavalry regiments and five infantry cohorts (one of them milliary). Detachments amounting to nearly 1,000 men from all these units were scattered at road stations along many of the highways of Britain. At Lincoln was Legion IV Hispana plus three auxiliary cavalry regiments and four auxiliary cohorts totalling 9,000 men. The auxiliary strength of the Roman Army in Britain was 17,000 men, compared to 22,000 legionaries. Earlier in the year Suetonius Paulinus had defeated the Ordovices in a fierce battle at Anglesey but his expedition had left the remainder of the Province denuded of troops, which encouraged Boadicea, Queen of the Iceni (who lived north of the Roman colony at Colchester with Norwich as their capital), to revolt against Roman rule. Aided by Coritani and Catuvellaun tribesmen, the rebellious host in an unruly mob descended on the

'Rameses' is not permitted to take any action until the frantic messenger arrives from the Division of Re; then he has to send out the Vizier, rally his forces and start counterattacking while awaiting assistance.

Certain problems of distance (ground scale) have to be considered: for instance, the 1½ (2.4km) miles that separated the Amon camp from the attacked Division of Re become 3ft (1m) on the wargames table, and the Vizier had to travel 4½ miles (7.2km) (9ft (3m) table distance) before reaching the detachment of Base Troops and a further ½ mile (800m) (1ft (30cm) table distance) to reach the Division of Ptah. Even if these forces react instantaneously, the Base Troops have to cover 4½ miles (7.2km) (9ft (3m)) and the Division of Ptah 5 miles (8km) (or 10ft (3m)) before reaching the scene of action. These troops consisted mainly of charioteers, but the Wargames Research Group Ancient Rules (the most widely used rules for this period) allow chariots only an 8in (20cm) move (18in (45cm) when charging to contact), at which rate it will take the Vizier 13½ game moves to reach the Base Troops and 15 game moves to reach the Division of Ptah. As these formations in their turn will take the same amount of time to reach the scene of action, Rameses' small force will have to face the might of the Hittite army for at least 27 game moves before the Base Troops arrive and 30 before the Division of Ptah comes to his aid!

As the majority of wargames are fairly conclusively settled in about a third of this time, the game is speeded up by introducing an illusory factor in the form of Chance Cards to determine the speed at which the Vizier and the reinforcing troops move. Chance Card No 1 allows the Vizier 4 game moves to reach the Base Troops and 5 to reach the Division of Ptah, and Chance Cards Nos 2 and 3 reduce his time further to 3 and 4, and 2 and 3, game moves respectively. A similar use of Chance Cards can then decide how long it takes the Base Troops and the Division of Ptah to reach the scene of action. A further set of these cards can be used to determine (a) whether the commanders of the Base Troops and the Division of Ptah are 'above average', so that they move quickly into action; (b) whether they are 'average', taking perhaps a game move to rally their troops and set them in motion; or (c) whether they are 'below average', and fail to move at all. Of course, the last contingency will completely alter the trend of the battle, probably reversing the result. All movements should be recorded on a Time Sheet, explained on page 15.

Commanders' Classification

Initially Rameses was outmanoeuvred, but he averted disaster by taking a firm grip on the battle. Mutallu, though beginning brilliantly, lost his advantage because he failed to exercise adequate personal command and control. Therefore, it would seem reasonable to class Rameses as an 'above average' commander and Mutallu as average. The Vizier, who played an important part in averting an Egyptian defeat, was above average.

Number and Quality of Men

The Egyptian army is said to have been 20,000 strong, a large portion being chariots divided into groups of 50, and the rest infantry. The Hittite army totalled 17,000, half its strength being 3-men chariots containing a driver and 2 spearmen. Neither side used cavalry. The Hittite attacking force consisted of 2,500 chariots containing 7,500 men, and about 3,500 infantry. The remaining 6,000 Hittite infantry were never committed.

The wargamer may find such large numbers of chariots difficult to amass, so these forces may have to be considerably scaled down, or cavalry used instead of chariots. The main Hittite striking force should be just over twice as strong as the destroyed Division of Re, and about twice the size of the Division of Amon. The Base Troops (part of the Division of Ptah) could be about a fifth of its strength, and the whole Division of Ptah about the same size as the Division of Amon.

Once they had lost the advantage of their initial surprise, the Hittites suffered from a technical inferiority in weapons, since the Egyptian chariots, their crews armed with the long-range composite bow, were far more effective than the Hittite charioteers, armed only with spears and javelins. The Egyptians also possessed elite troops in the Na'Arun Canaanites. In the New Kingdom of Egypt (from 1580BC onwards) charioteers and bowmen wore a helmet, and mail armour made of rectangular scales of metal. Infantrymen carried a small shield and used spears as their basic weapon. The composite bow, supplied to both charioteers and infantrymen, could penetrate the armour of the time at a reasonably short range.

Morale

At the outset the Hittite morale was first class, whereas that of their first opponents, the Division of Re, must have fallen to second or even third class by the surprise flank attack. As the battle progressed, the Hittite morale faltered, whereas that of the Egyptian Divisions of Amon and Ptah and the Base Troops remained consistently high.

Style of Fighting

Egyptian chariots operated as a unit, charging in a solid mass to drive the enemy before them, and manoeuvring so as never to allow themselves to be individually surrounded. The Hittite method of chariot attack was to charge the enemy pell-mell, and, if the fighting became too congested for manoeuvre, dismount the driver and turn him into an extra spearman. The spearman on both sides formed in large groups, rather like a primitive phalanx, while the lightly armed archers and slingers swarmed out on the flanks and skirmished in front.

Terrain

All that is required is a gently undulating area dotted with patches of scrub and perhaps small clumps of trees.

undefended town of Camulodonum (Colchester) and razed it to the ground. The IX Legion, marching from Lindum (Lincoln) to quell the revolt, were overwhelmed by sheer numbers and almost wiped out while Paenius Posthumus, the commander of the II Legion, judged it wisest to remain in the protection of his camp at Gloucester. After destroying Londinium (London) and slaughtering its inhabitants, the horde of Britons were brought to bay by Suetonius Paulinus with the XIV and the XX Legions on a carefully chosen ridge with unturnable flanks, said to be near Staines in Middlesex. The legionaries, six deep, formed a 1,500yd (1350m) crescentic line, with cavalry on the wings and with the bodyguard cohort of Suetonius Paulinus in the rear, across the causeway, as the only available reserve. The auxiliary infantry were posted, as skirmishers, in a long straight line some 1,000yds (900m) forward. The Britons poured on to the battlefield like flood-water through a broken dyke and after taking casualties from the auxiliary Bracarii slingers and Thracian archers plus short sharp cavalry charges into their flanks, they flung themselves in a headlong charge on the skirmishers who were recalled to the flanks by trumpet. The two Legions, swords held low and shields up, went forward down the slope in a steady controlled run; at twenty paces' range the front rank flung their pila. Then both Legions formed into a series of 'cohort wedges', with one century at the apex, two more echeloned rearwards on each flank and the sixth at the base of the wedge to support and reinforce where required. The British rush was brought to a *continued*

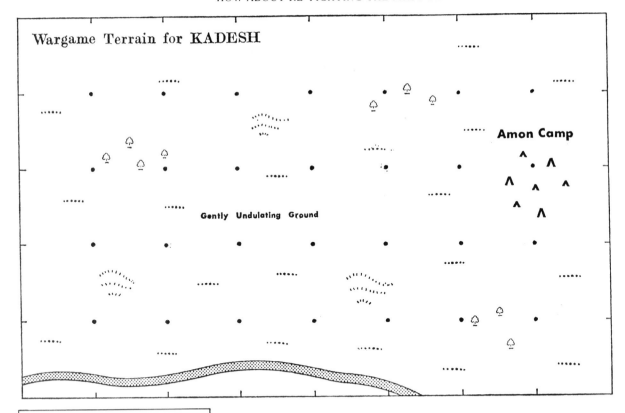

Wargame Terrain for KADESH

Amon Camp

Gently Undulating Ground

standstill and only a 1,000 Britons could engage the Romans on the unflankable narrow front. Soon the Britons began to give way until the wedges had flattened and the Roman battle-line was once again a straight line; then the auxiliary cavalry charged down the slopes to smash into both British flanks and dissolve them. The British front broke into a rabble fleeing in all directions to be trapped in a 'killing-ground' bounded by Romans and a barrier formed of their own carts and wagons. It is reported that 80,000 Britons died; the Romans lost about 400.

Military Possibilities

1. Rameses does not allow his columns to straggle, but waits at the ford of Shabtuna until each division has crossed, and then moves forward as a united force. This will probably save the Division of Re from surprise attack.

2. Though surprised, the Division of Re is not routed.

3. After routing Re and descending on the Amon camp, the Hittites press their advantage instead of halting to plunder.

4. Rameses and his Division of Amon are completely dispersed, unable to rally and counterattack.

5. Either the Vizier is not sent or he does not reach the on-coming Egyptian columns.

6. The Base Troops and/or the Division of Ptah do not arrive in time to aid Rameses. The classification of their commanders can have a bearing on this.

7. Instead of remaining on the far side of the river, Mutallu, in the later stages of the battle, brings over the remaining 6,000 Hittite infantry; or he exploits to the full his initial success by immediately committing all his chariots, and despatching the mass of his infantry by the same ford or round to the north of the city to back them up.

In some ways Kadesh resembles Salamanca, with the Egyptians filling the French role but doing what the latter failed to do by turning defeat into victory.

11
EARLY WHEELED WARFARE –
The Chariots of the Ancients

Until about 700BC, the elite striking force of the armies of antiquity were usually chariots, small armoured carts, often with sharp blades projecting from their whirring wheels, and drawn by armoured horses. Royalty, Great Nobles and leaders rode to battle in chariots and at times fought from them, although usually dismounted for actual hand-to-hand fighting.

There was nothing new in the concept of combat vehicles. The use of chariots and wheeled battle wagons goes back to the dawn of recorded history. The fast two wheeled chariots of the Assyrians, used more as fighting platforms than as weapon carriers, dominated wars between the years 1100 to 670BC. It was in Sumer (the lower part of Mesopotamia) that fighting turned from a sporadic into a systematic activity. Civilisation first arose in this area in about 3000BC; later, the Sumerian cities began to fight each other and so the institution of war was bestowed upon mankind. Paintings on pottery reveal that the chariot was used by the Sumerians, probably more as a means of transport to carry a chief or champion into battle than as an actual weapon.

Ancient Egyptian chariot,
c1215BC

Drawn by onagers (a species of ass) they had solid wheels and were clumsy and slow. Sometimes a leader, if finding himself hard pressed or out-marched, would flee the field by jumping into a conveniently placed chariot and hastily leaving the scene of action.

The Egyptians and the Greeks used light chariots drawn by two horses with a crew of driver and bowman. Not so extensively used by the Greeks, chariots were the backbone of the Ancient Egyptian army. The Assyrians, Gauls and Ancient Britons used heavy chariots drawn by two or four horses and carrying a driver with two men.

At about the same time as the Egyptians flourished, the new nation of Assyria was coming to the fore in the East, inheriting and extending the ancient civilisations of Sumer and Babylon in the fertile country near the head waters of the Tigris. The main striking force of the Assyrian Army was the corps of horse-drawn, two-wheeled chariots, whose role was to smash their way through the ranks of enemy infantry. Like their contemporaries, the Assyrians used chariots in simple, brute force, but in larger numbers, with more determination, and in closer co-ordination with archers, spearmen and cavalry. Chariot crews formed the elite of the army. The chariots were of two types – a light

An ancient Indian army, with chariots and elephants. (Museum Miniatures)

Ancient Egyptian chariot c1215BC (Yadin)

two-man vehicle bearing an archer and a charioteer and a heavier chariot carrying four men – archer, charioteer and two shield bearers. These vehicles had no springs which must have made the firing of a bow from them a difficult business so that it is quite possible that at times the chariot halted and the bowmen either fired from a stationary platform or descended and fired their missiles from the ground. Egyptian chariots had a 'floor' of crossed leather thongs, thereby giving some measure of 'springing' for the occupants. It could be that the Assyrians used something similar – they had been in contact with the Egyptians for some long time. While the archers were in action their attendant shield bearer warded off enemy missiles. Reliefs of the time show the chariot horses to be wearing a type of barding or protective armour, probably of some heavy quilted material. The chariot crew wore short sleeveless coats of scale armour made of small overlapping plates sewn on to a cloth backing. They carried short swords slung by a strap on the left side and small round shields; they wore the typical high peaked helmet with small lappets over the ears.

Reliefs and sculptures show that the early cavalry horses had equipment almost identical to that of the chariot horses and show horse-archers being attended by an unarmed man, which leads to the theory that perhaps these archers were originally charioteers who had unhitched the horses and ridden off when the ground was unsuitable for the use of chariots.

Around 2000BC a new civilisation had grown up in the Aegean. Originally created by the Cretans, it flourished under the Myceneans or Greeks of the mainland, whom Homer later immortalised in his epics. The Achaeans, northern invaders, conquered and finally mingled with the original inhabitants of this Minoan or Aegean civilisation. From the mixing of the

An ancient Briton

The Bowmen of Agincourt
A single volley by the archers at
Agincourt amounted to about
5,000 shafts, each weighing 2.5
oz (70g), a total weight of 781lb
(355kg); the total weight of all
the arrows that fell upon the
French on that day, assuming
that each archer loosed all his
stock of 48 shafts, was nearly 17
tons. In this battle there is little
doubt that by far the greater
proportion of the 6–7,000 French
dead lost their lives in close-
quarter melee-fighting – by
concussion, cuts or stab wounds;
yet, English archers probably
loosed 240,000 shafts during the
course of the battle! Generously
allowing that they killed 25
percent of the total French dead
– it took 150 arrows to kill one
man! However, that is far from
the true effect and long-term
potential of the arrow-hails that
darkened the skies during the
Hundred Years' War.
continued

two races and cultures came Homer's 'Heroic Age', placed from
about 1500BC to 1100 or 1000BC. It was probably during this period that the struggle occurred between the Achaeans with their
allies and the rulers of the lands around Troy, immortalised in
Homer's poems 'The Iliad' and 'The Odyssey'. Despite the beauty
of this poetry, warfare of the time was almost devoid of planned
manoeuvring or real leadership. The 'Heroes' rushed into battle
in their chariots followed by the infantry 'shield to shield, shoulder to shoulder' to take part in a vast sprawling melée where
spears, rocks, swords and daggers took their close-quarters toll.

The most interesting contest of the period was the Trojan War
which lasted from about 1194 to 1184BC, when Agamemnon led
his armies of chariots and infantrymen into Asia Minor to put
an end to the menace of the Trojans who had established themselves in Troy, their fortress kingdom. Homer tells of chariot
lines meeting each other with great shocks, vast melees between
on-rushing infantrymen and hails of javelins as soon as chariots
came within range, when the occupants leapt from their carts to
attack the enemy on foot.

In 55BC, Julius Caesar, with a force of two legions plus auxiliary
cavalry, crossed the straits of Dover and invaded Britain, pushing only 10 miles (16km) inland before rough weather damaged
his ships so forcing him to retreat. Next year he returned with
five legions and drove across the Thames to penetrate into what
is now Hertfordshire. Caesar wrote about his battles with the
British, dwelling at some length on the chariots which came to
meet him the moment he landed.

'The Britons begin by driving their chariots all over the field
hurling javelins . . . the horses and the noise . . . are sufficient to
throw their opponents ranks into disorder. Then, after making
their way between the squadrons of their own cavalry, they jump
down from the chariot and engage on foot. In the meantime their
charioteers retire a short distance from the battle so that they
are handy in case of retreat. Thus they combine the mobility of
cavalry with the staying power of infantry.'

The chariot used by the British was open in front with
two spirited native ponies harnessed to a broad, flat pole. Two
fighters rode in the chariot including the noble who drove and
commanded it – sometimes one of them would run out along
the pole to fling javelins from between the horse's heads. In
action, the tactics were to race the chariots in a column straight
at the opposing foot soldiers, sending a shower of short javelins
amongst them if they broke. If they did not break, the chariots
would wheel off along the front with the fighters flinging their
javelins as they passed – sometimes they would circle the enemy
formation throwing in another volley of javelins at their rear.
After several attacks of this kind, the chariot fighter dismounted
and went in with the infantry. Caesar was impressed by the
way British cavalry and chariots retired into the woods when
they were repulsed, and by 54BC, when he returned to Britain,
Caesar had worked out the correct tactics to use against British

Assyrian Cavalry and Archers

chariots. He decided to attack them on sight before they could draw up into a defensive formation. Throughout this campaign Cassivellaunus, the British king, constantly harried the Roman columns with his chariots so that the legionary columns had to be protected by outlying cavalry. Much as they harassed the Roman auxiliary cavalry and Caesar's foraging parties, time proved that chariots could not break the solid ranks of legionary veterans.

After the death of Julius Caesar the Britons were left alone for almost a century until in AD43 Claudius came with an invading force. Claudius believed the British warrior, man for man, to be at least equal to the Roman but considered that his methods of fighting and his arms were more suited to single combat. The Briton fought with a broad-sword, an unwieldly weapon at close quarters, and his leather buckler was too small to provide much protection against the powerful pilum. It was during this campaign in September AD43 that the Romans won an extraordinary victory in the misty marshes near Romford in Essex. When the first column of British chariots rattled down upon the enemy

English longbow men of the Hundred Years' War formed-up and fought in a disciplined solid mass, a wedge-shaped formation (Harrow or Herce), allowing the maximum number of bows to be brought into action in converging volleys. From this forward-pointing, bastion-like formation, they shot in volleys under direction of marshals or the master-bowmen. It must have been as horrifying as the first sight of a war-elephant, the deadly effectiveness of this with 5,000 skilled and experienced bowmen standing shoulder-to-shoulder, shooting collectively with each man loosing off at least five shots per minute, filling the air with thousands of arrows in flight at once followed by similar waves at twelve second intervals, the first wave arriving on the target as the last wave left the longbow. The English archer could maintain his rate of firing as long as he had arrows.

This formation was gloriously victorious at Crecy, Poitiers and Agincourt. Smaller forces may have formed-up their archers in front of the men-at-arms in five groups each of five men.

*The first organised soldiers –
Sumerians 1500BC*

their horses were thrown into confusion by a foul smell that filled the air – it came from a line of camels, creatures completely unknown to the Britons. With the leading chariots milling round in confusion, their horses practically out of control, the legionaries showered them with pila. Other chariot attacks were thrown back by showers of blazing balls of pitch thrown from war engines, together with hails of missiles from the auxiliary slingers. Trip ropes had been placed in position during the night and the horses and chariots piled up on them. To add to the confusion the Britons found themselves attacked by coloured auxiliaries whose unfamiliarity caused them to be treated as black devils. It is said that, on at least one other occasion, the Romans brought elephants to Britain, with no doubt immense moral effect. Claudius used them for effect but never in action against the Britons.

As time passed, except in India and Persia, the chariot lost much of its terror for disciplined and manoeuvrable infantry, and was no longer a principal weapon of battle. This was demonstrated at the Battle of Chaeronea in 86BC, when Sulla, with about 30,000 men, marched into Boeotia to battle with Archelaus's army of 110,000 men and 90 chariots. In the first-ever offensive use of field fortifications, Sulla had entrenchments built to protect his flanks from being enveloped by the Mithridatic-Greek cavalry, and erected palisades along his front to protect it from chariots. After Sulla had repulsed a cavalry charge with his legions in square, the chariots attacked but the horses, maddened by Roman arrows and javelins, dashed back through the phalanx, throwing it into confusion. Sulla immediately launched a massed counter-attack and drove the foe from the field.

Here is a basic set of rules for controlling chariots on the wargames table; they do not differentiate between heavy and light vehicles and may be adapted to personal taste – in wargaming, one learns a lot from experience.

1. *Chariots against infantry.* If infantry are in double ranks or more, chariots throw dice and need 4, 5 or 6 to charge home, one die being tossed for each chariot. If chariot does not charge home, throw further die: 1, 2 or 3 chariot swerves to right and completes its move distance, 4, 5 or 6 chariot will swerve to left and continue its full move distance. If the chariot charges home, the infantry will throw dice and need 3, 4, 5 or 6 to stand. If they run, all in the path of the chariot will need dice thrown and only those who throw a 6 will escape. If infantry stand, chariot fights number of men equal to number of horses and men in chariot, one die throw per group. If chariot is beaten, throw dice for horses and occupants of chariot individually – 5 or 6 to save – if horse is killed the chariot is useless but men may dismount and fight on foot.

2. *Chariots against cavalry.* Throw a die for each side, if cavalry throw a 1 they turn and bolt and need to throw a 5 or 6 individually to survive. If chariot throws a 1, then throw dice individually for each chariot – 1 overturns, 2 bolts full move, 3

or 4 swerves full move right, 5 or 6 swerves full move left. If they engage, each chariot fights two cavalry.

3. *Chariots against chariots.* Both sides toss dice, if one throws 1 then that side swerves and no combat. Otherwise, throw dice first for each chariot on both sides, any throwing 5 or 6 collides and overturns, after that it is chariot against chariot, straight dice throw, highest score wins.

4. *Casualties from missile fire.* Number chariots and then throw die to see which chariot is hit and how many times. Then throw again to see whether it is the driver, soldiers or a horse that is affected. If drivers or soldiers, they can be saved by 5 or 6. If horse or driver is killed, chariot overturns and kills all occupants.

5. *Chariot against elephant.* Elephants cannot be attacked by chariots but an elephant can trample two chariots in one move provided they are within his move range.

On the wargames table the handling of chariots depends upon whether they are heavy or light – former being essentially a shock weapon, the latter a weapon of manoeuvre. The main effect of the heavy chariot will be achieved by the weight and shock-power of the horses and the chariot itself, both almost certainly armoured. Unless suffering too badly from enemy missiles during their approach, they can burst upon infantry formations and send them reeling back; they are also effective against heavy cavalry. The light chariot is used for harassing the enemy, moving rapidly around his formations, pouring in missile fire. Avoid close contact and shock-action, keeping the chariots on the flanks or to threaten those of the enemy. If rules permit, chariots can be employed to transport infantry, picking them up, so many to a chariot, and hastening them forward to a key position or threatened point. Use chariots in mass, not in penny packets.

Assyrian Cavalry

An exquisite model of an Egyptian chariot (Rose Models)

WEATHER IN WARGAMING

The effect of weather upon wargaming armies was sometimes quite profound, yet few Wargamers have any rules to take its effects into consideration. At Crecy, in 1346, the English archers hastily unstrung their bows and stowed the bowstrings away in a safe place when a sudden rainstorm fell upon them. The mercenary Genoese crossbowmen employed by the French were almost completely useless when the battle started soon afterwards, their crossbows having been sadly affected by the sudden deluge.

There are numerous instances recorded during the Napoleonic Wars when weather, particularly rain, dramatically affected the combatants. It was a period when the infantry square was seldom broken by a cavalry charge but heavy rain could alter this. Muskets cannot fire and cavalry became supreme as at Dresden in 1813 when a French lancer charge easily crushed an Austrian square because the Austrian muskets could not fire owing to the rain. In the same battle, an Austrian square quickly surrendered when threatened by cavalry and Horse artillery cannister in the rain, the Austrians realising that they were defenceless. Without fire-power even the renowned British Infantry of the period could be badly hurt and that Albuera in 1811, a French lancer charge in the rain

Napoleon's retreat from Moscow
1812. The French rearguard

destroyed three British Infantry battalions in three minutes. When the muskets cannot fire bayonet charges become formidable indeed and at Ligny in 1815, the French Guard Infantry successfully bayonet charged the Prussians in the rain. With muskets unable to fire, a French column charge was as effective as that of a Greek phalanx. At the Katzback in 1813, the French had the tables turned upon them when the Prussians copied the French style of column charge and successfully attacked with the bayonet in the rain.

Every military historian is conscious of the manner in which Napoleon's artillery and cavalry were hampered by the mud at Waterloo. It would not be unrealistic to say that the mud might have been responsible for the British victory because Napoleon did not attack until mid-day, hoping that the ground would dry out. Had it not been wet and he had been able to hammer the British for an extra four to six hours before the Prussians arrived, the battle might well have taken a very different turn. Nor does it take very much imagination to picture the devastating effect that the snow must have had upon the French during the Russian campaign of 1812. There are some Germans living today who will also recall with a shudder the similar effects of snow and mud upon Hitler's forces during the great battles in Russia of World War II. Napoleon's hard won victory at the battle of Eylau in February 1807 was also much affected by blinding snow storms.

Mist and fog can be a very decisive factor in war. At Fuentes d'Onoro in 1811, the British Allied right

wing outposts were surprised by French troops looming out of the early morning fog and were attacked before a defence could be formed. Auster-erlitz in 1805 and Inkerman, during the Crimean War in 1954 were also stern battles in which fog played a big part. Many years earlier, in 1471, thick fog caused the battle of Barnet, during the Wars of the Roses, to become a series of groping mis-calculations.

The wind can also have an effect upon tactics. When the Yorkists marched out to give battle against the Lancastrians on 29 March, 1461, at Towton they had the wind at their backs and were fighting during a heavy snow-storm. Aided by the wind, the arrows of the Yorkist archers had a greater range than those of the Lancastrians; the Yorkist archers fired a few volleys to draw the Lancastrian fire and then withdrew, their move-ment being unseen owing to the blinding snow. The Lancastrians sent volley after volley of arrows until the quivers were empty without a Yorkist falling. In the snowy field the uselessly expended shafts were sticking up like porcupine quills. The Yorkist archers then advanced and with the wind behind them slaughtered their opponents, even gathering up their own wasted shafts and using them against the Lancastrians. Even the absence of wind can be put to good use – during the Napoleonic Wars it was common for French skirmishers to fire furious-ly so as to create a smoke screen. At Friedland in 1807, Lannes was able to conceal the weakness of his force of only 10,000 men from Beningsen who had 46,000. Later in the day, Ney attacked on the right completely covered by smoke so that the Russian cavalry counter-charge ran down their own men whom they were unable to see.

Weather rules for wargaming must be designed to give both strategical and tactical effects upon operations. Not only should the actual battling on the table-top be influenced by the vagaries of the climate but similarly, the moving on the map that is essential to all campaigns must also be affected by the weather.

First divide the year into the four seasons of spring, summer, autumn and winter and then list the weather that is possible in each season.

SPRING (March, April and May)
Torrential rain, Light rain, Fog, Mist, High Wind, Sunny, Dull, Thunderstorm, Average

SUMMER (June, July and August)
Torrential rain, Light rain, Mist, Fog, Intense heat, Bright and sunny, Dull, Thunderstorm, Average day

AUTUMN (September, October)
Torrential rain, Light rain, Mist, Fog, Snow, Sunny, Dull, High wind, Intense heat, Average

WINTER (November, December, January and February)
Torrential rain, Light rain, Snow, Blizzard, Mist, Fog, Dull, Bright, High wind, Average

Prepare a set of cards for each season, bearing in mind that the chances of these various types of weather vary greatly according to the season. As an example, when making out the summer cards – have one card for thunderstorms, one for torrential rain, one for intense heat, three for average and two for each of the others. In this way, the chances of a thunderstorm would only be 1 in 16, whereas the chances of average weather would be 3 in 16 and since sunny or dull weather have little or no effect, the chance of reasonable weather in the summer is 7 out of 14.

To use these cards, one should be drawn before each map-move or before battle at the actual start of the battle. It could be considered that in Spring and Autumn, when the weather is likely to be changeable that this procedure of drawing a card should take place. But in Summer and Winter, when the weather is possibly more settled, it may be advisable to merely dice each time to see whether there is any change in the weather and only draw a weather card if there is. When considering weather during the map-moves of a campaign, it is a good idea to decide on a state of weather by drawing a card and then throw a dice for each map 'day' to see if the weather has changed or remained constant.

When considering the strategical and/or tactical effects of the weather on operations it is necessary to discard average weather, dull weather and bright sunny weather because none of them have any effect whatsoever. It will also be necessary when considering tactical influences, to assess the possible effect of the weather upon the specific period in which one is fighting. It would be unreasonable to assume that modern operations would be affected quite so severely as those of Napoleonic or Ancient times.

To represent the practical effects of weather on actual tabletop battle operations, the following can be used as a rough guide.

Torrential Rain On the first day, the rate of movement of troops and vehicles is halved. Houses on fire will burn at only half normal rate. On the second day of such rain, movement is only possible

Cataphracti in the snow (Figures by Miniature Figurines – courtesy of Bob Douglas)

on roads and at half-rate; creeks and rivers become uncrossable; dice to see if bridges are down (scores under 3). Conditions remain thus for the day after the rain ceases. At all times during rain, bows, crossbows and muskets are affected – being unable to fire under conditions of torrential rain, and at a decreased rate under normal wet conditions.

Snow All movement is halved. Visibility limited to an agreed distance. Effectiveness and rate-of-fire of bows and muskets decreased by half, and the effective range of all weapons reduced by 6 inches. Under blizzard conditions, everything stops.

Fog No action possible or visibility down to an agreed distance. Aimed missile fire drastically reduced in range and frequency. Less drastic measures for mist but it has its effects.

Intense Heat Movement halved. Marshes, rivers etc dry up or become more crossable. Woods, houses etc more likely to catch on fire.

High Wind Fire from bows and crossbows less effective against the wind, but better with it. Dust storms may affect visibility. Fired buildings and woods will burn twice as quickly and fire will spread.

Thunderstorm No archery possible during storms. Rules can decide if lightning strikes and causes ammunition explosions etc. Fear-effect may be present on Ancient or native troops.

The continuity of weather may be represented by additionally picking a weather-card and, by dice throw, decide which card prevails.

A French rearguard under attack by Cossacks, during the retreat from Moscow in 1812

12
WELLINGTON IN THE PENINSULAR 1808–1814
Britain's Most Stirring Overseas Campaign!

There can be few wargaming periods offering more colour, interest and general satisfaction than the Peninsular War of 1808–1814, when the Duke of Wellington hall-marked his military career in a series of brilliantly conceived victories by the tightly-knit and highly professional Anglo–Portuguese army he led. Emblazoned on the colours of many British infantry regiments, the names of the battles ring proudly round the halls of military history – Vimiero, Busaco, Fuentes D'Onoro, Salamanca, Vittoria, Nive, Nivelle – still laying almost unaltered 180 years on. Extensively researched and walked on numerous occasions by the author, their contours and varied topographical features laid out on Southampton wargames tables have become familiar.

The British army of this period was by no means typical, consisting of a small, well-trained and disciplined professional force as opposed to the French and other Continental armies which were huge masses of citizen-soldier volunteers and conscripts. As a result, the British infantryman was more adept with his weapons, able to carry out more complex manoeuvres and withstand greater punishment than his Continental contemporaries. For much of the war it consisted of 7 Divisions, with a Light and a Cavalry division under separate command. Usually, British divisions consisted of about 6,000 men in two British brigades and one Portuguese brigade, each of three battalions, with a battalion of Cacadores (light infantry armed with rifles) to each Portuguese brigade. Every infantry brigade was given an extra company of light riflemen to reinforce the three light companies which were now standard in the British brigade.

To enable wargamers to accurately scale-down their forces, the British organisation in the Peninsular was:

A French Hussar of the Napoleonic period; the 3rd Hussars of Lamotte's Brigade braved the guns of Almeida to sweep across and onto the flank of the Light Division

INFANTRY	50 men = 1 company
	10 companies = 1 battalion (which included 1 grenadier and 1 light company)
	2 battalions = 1 regiment (second battalion usually in England)
	3 battalions = 1 brigade (plus 1 company of rifles)
	3 brigades = 1 division
CAVALRY	120 men = 1 squadron
	3 squadrons = 1 regiment (2 squadrons for Household Cavalry)
	3 regiments = 1 brigade

A British rifleman of the 95th; Peninsular War Period

The cavalry were Wellesley's weakest arm and invariably he endeavoured to select battlefields that did not favour horsemen – fortunately, the Spanish countryside made this possible. Outnumbered, it was not possible to match them against French horsemen in the big charges that the French were able to perform so expertly. British cavalry swords were made blunt by the metal scabbards in which they were encased whereas the French used wooden sheaths.

An important part of the British force was the King's German Legion, formed from the remnants of the Hanoverian Electoral Army, with a peak strength of 2 horse and 4 foot batteries; 2 regiments of dragoons; 2 regiments of hussars; 2 light and 10 line battalions, each of 10 companies of 119 men. The cavalry regiments consisted of 4 squadrons, each of two 90-men troops. The King's German Legion light infantry companies had ten riflemen each, so that their 100 per battalion was considerably more than a company per brigade. The artillery battery consisted of 6 guns and 175 artillerymen, 8 ammunition and 2 baggage waggons.

The cream of the British Peninsular army was the Light Division, with their essential role of countering the French Tirailleurs who skirmished ahead of their columns. Thoroughly trained, the British light infantry were able to operate in close order or as individuals in a manner far superior to the average French tirailleur who frequently was only another conscript in a different uniform stuck out in front of the columns. A scattered rifle regiment could quickly re-form to present a front to deliver the equivalent of a musketry-volley. This is all explained at greater length in Chapter 8.

In 1810, the British Light Division under General Robert Craufurd consisted of:

'A' Brigade – 1st Bn 43rd; 4 companies 1st Bn 95th and 1st Bn Portuguese Cacadores.
'B' Brigade – 1/52nd; 4 companies of the 1/95th and the 3rd Cacadores.

Two battalions of the 95th, the 5th battalion of the 60th and the Cacadores of the Light Division were all equipped with the Baker rifle, which could be loaded and fired, from a lying position if necessary, in 30 seconds but under battlefield conditions once per minute with an average misfire of one in every 13 rounds.

French columns often outnumbered British lines but were invariably repulsed in a manner that astonished the French, aware that their massed formations, tirailleurs and artillery had defeated the best that Europe could throw against them. The musketry advantage that the infantry line had over a column was usually outweighed because they were so weakened by the fire of light infantry and close-range artillery that they were overcome by the shock-effect of the columns. This was all altered by Wellesley's feat of countering the tirailleurs with his own riflemen, while the artillery accompanying the columns were unable

The Great Northern War (1700–1718)

Charles XII of Sweden came to the throne in 1697, two years later Denmark, Poland and Russia formed a coalition to take advantage of the 17 year old king. Brilliantly leading his well organised armies, King Charles defeated Denmark in a two weeks' campaign and knocked them out of the war. Next he marched to fight Poland and Russia and gained a great victory at Narva where the redoubtable resistance of the Russians persuaded him to postpone an invasion of Russia for an attack on Poland, the more dangerous of his foes. After occupying Warsaw Charles campaigned constantly from 1703 to 1706, easily defeating Saxons, Poles and Russians in a number of battles. In 1708, at the head of 20,000 infantry and 24,000 cavalry, Charles of Sweden invaded Russia. Like Napoleon and Hitler, the Russian winter defeated the Swedish King, and in *continued*

to deliver supporting fire because horses and artillerymen were casualties before they could come into action. Like King Edward, the Black Prince and Henry V during the Hundred Years War, before he would give battle Wellesley invariably sought a position, preferably on back-slopes, that best suited his troops. The greatest single virtue of these defensive positions was that they always permitted a rapid and continuous flow of reinforcement to threatened sections, often on quickly constructed roads running behind the position.

When French columns came under fire from British lines, they were forced to abandon their attack and endeavour to hold their ground. With time to adjust and convert the action into a static fire-fight (as at Albuera), the resulting action was in 18th-century style, usually lasting about twenty minutes before one side broke, often because its antagonist showed signs of coming forward, as when reinforced. British infantry attacks on columns were made immediately after their first volleys, which were rarely the oft-mentioned 'rolling half company volleys for long periods', because the musketry was a preliminary to the possible use of the bayonet, which, however, was rarely used as formations broke and fled before contact. When wargaming, there should be more stress on morale than on formation in this aspect, although formations were important in fire-fights, as indicated by French column and British line.

the spring of 1710 his much depleted army faced the vastly superior Russian force at Poltava. The Swedish columns advanced with all their old courage but were decimated by flanking fire from a series of Russian redoubts and finally the immense Russian superiority of artillery almost annihilated the Swedish army. It was a battle that had far reaching results, destroying forever the power of Sweden and convincing the rest of the Continent that the Russians now had to be taken into consideration. After Poltava, the Prussians, the Danes and the Russians all combined to bring Sweden to the ground and the nation finally had to sign a humiliating peace.

British riflemen and French Tirailleurs exchange fire over the Coa

French infantry of the Napoleonic period similar to those at the Coa

French Napoleonic cavalry similar to those who did so much damage at the Coa

Opposite:
An impressive Napoleonic battle using AIRFIX HO/OO scale plastic figures in a refight of the Battle of Vimiero in Portugal, 21 August 1808. (John Tuckey Collection. Photo by Richard Ellis, courtesy of Miniature Wargames)

The British system of halting a column attack with volley-fire and then immediately charging the demoralised survivors can be simulated by splitting each move into two or three parts so that one player takes a moving move, while his opponent takes a simultaneous firing move. The 'moving' move is a three-part move and allows charges, etc, while the 'firing' move is a two-part move that permits the unit to wheel or change formation and then fire.

It required steady, disciplined troops to operate in the British manner and Blucher remarked that his troops, even in three-deep lines, did not have the discipline, training and individual stability to hold such formations, so they did not use them. The British (and later the Portuguese) soldier in the Peninsular was a professional with superior qualities over a conscript, enabling British infantry to deploy in line and so employ maximum firepower.

Throughout the Peninsular War the Spanish troops, with the exception of the experienced battle-hardened men of Morillo's and Longa's units, were unreliable and untrustworthy; their morale was unstable and, after an opening volley that was misloaded and unco-ordinated and unaimed, there was usually a wholesale flight. Spanish cavalry would only go into action against disordered infantry; but their artillery, when not influenced by their countrymen fleeing round them, frequently performed heroically and well.

When attempting to re-create historical battles on the wargames table it is often found that the sheer extent of the real-life field has to be markedly condensed, so that any attempts at authentic scales are doomed from the start. However, there is one classic Peninsular action that embodies every typical aspect of the warfare of the period, besides being fought over a compact terrain memorable to the author who has 'walked' it at least four times, which can comfortably be laid on an average-sized wargames tables. Showing the incomparable Light Division at their scintillating best, Craufurd's Combat on the Coa should be in every wargamer's repertoire – if it isn't, now's the chance to become acquainted with it!

In July 1810 General Craufurd's Light Division took up position on rugged ground descending from the glacis of the fortified Spanish-Portuguese border town of Almeida, down to the River Coa where their right flank rested, the left on a ruined windmill within range of guns on Almeida's ramparts. Here, just after dawn on 24 July 1810, Craufurd was attacked by Ney's far larger force. There can be few engagements in the whole history of warfare that lend themselves more to tabletop re-enactment than this fight on the Coa, made incomparably better if one has 'walked' the field, noted its features and area where most action occurred, and stood on the bridge that was packed to parapets with bodies of French dead. When re-constructing an actual battle, with terrain faithfully scaled and forces positioned as on the fateful day, wargamers can fight it as they see it, when hindsight

Notes on Artillery in the Time of Napoleon

It is easy enough to work out the effective range of the artillery of those days by studying the maps of the various battles and seeing where the batteries were placed. Even so, it may be just as well to point out that contemporary French theorists considered 1,100yd (1,000cm) as the maximum effective range in war.

A 12pdr given 6° of elevation could carry 1,900yd (1,700m), and the smaller field guns, 8pdrs and 4pdrs, could carry almost as far, but it was considered that beyond 1100yd (1,000m), 'objects became too indistinct for fire to be at all accurate' (*The Background of Napoleonic Warfare*, by Robert S. Quimby, p.294, quoting Du Teil). The range of howitzers was 1,000yd (900m) at 4° elevation, and this provided four or five ricochets on favourable ground. It was desirable to place artillery in such a way that a grazing fire would be possible: a slight elevation was better than a great one, which would produce a plunging fire and cause the shots to bury themselves, thus much reducing the effect of solid shot.

It is worth noting that 8 men could handle a 4pdr, while it took 11 to move an 8pdr and 15 to manhandle a 12pdr.

Before the reforms of Gribeauval, the equipment of the French Army was very much heavier and field artillery was not easy to handle on the battlefield. It was important to lighten the pieces so that the horses could be left under cover. By Napoleon's time a 4pdr could easily be hauled by 4 or even 3 horses.

It is thought that at Waterloo the British artillery expended 9,467 rounds. Since an artillery waggon drawn by 4 horses carried about 100 rounds, it will
continued

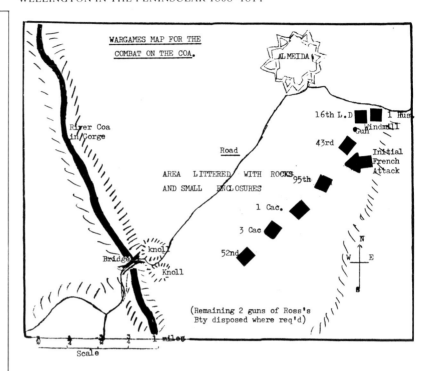

WARGAMES MAP FOR THE COMBAT ON THE COA.

ALMEIDA

River Coa in Gorge

Road

16th L.D   1 Hus
Gun   Windmill

43rd   Initial French Attack

AREA LITTERED WITH ROCKS
AND SMALL ENCLOSURES   95th

1 Cac.

3 Cac

52nd

Bridge   knoll

Knoll

(Remaining 2 guns of Ross's Bty disposed where req'd)

N   W   E   S

Scale   ¼   ½   ¾   1 mile

usually makes it a very different conflict, authentic in name, terrain and opposing forces – but little else. Uniquely, the Combat on the Coa lends itself almost ideally for fighting as one pleases, yet still retaining more than a semblance of reality.

Beyond Almeida the Plains of Leon stretch in a low rolling treeless expanse that contrasts strangely with the wild, rocky and picturesque area on the westward side of the town, where it plunges down sharply to the gorge of the River Coa, only a mile and a half away. The upper part is broken into small fields by walled vineyards, high walls and little enclosures; the last part down to the river is rock-strewn and covered with heather and broom. Descending steeply down to the bridge from Almeida was a slippery partly-paved road enclosed by high banks and stone walls; nearing the river it turned sharply upstream before doubling back to the bridge, making difficult hasty progress of guns and horses. Anyone descending to the river will not see the bridge until reaching water-level, as it is masked by a pine-tree covered knoll which covers the bridge from attack or fire from above. The bridge, 70yd (63m) long with a curious twist in its middle, crosses the river in a narrow rocky gorge where the water runs fiercely. A difficult terrain to fight over for both attacker and defender was this stage set by nature for Craufurd's Combat on the Coa, as sharp a fight on its own scale as any seen in the entire course of the Peninsular War.

With his left on the windmill Craufurd had his line formed across the slope of the hill facing roughly East, his left 500–700yd (450–630m) south of Almeida, where half a company of the 52nd with two of Ross's guns held the windmill; then came the 43rd; followed in succession by the 95th; 1st Cacadores; 3rd Cacadores, and the remainder of the 52nd nearest to the river – the whole line in convex form covering a front of about 1½ miles (2.4km), with cavalry picquets dispersed out in front. Craufurd's position on the exposed ridge was not a good one, if attacked, the river-defile and only a single narrow bridge to his rear meant he could either be hustled down to the bridge and forced to pass the whole Light Division across it in dangerous haste, or thrown into Almeida to become part of the trapped garrison. Wellington was conscious of the risk, and repeatedly warned Craufurd against becoming involved in the area. If attacked on the Coa, Craufurd intended to conduct a skilful rearguard action, not going until pushed; in all fairness, walking the ground certainly leads to the belief that it would be hard to find a more suitable battlefield for a detaining force – providing the enemy were in no more than moderate strength. There are numerous successive points to be held one after the other, and the many stone-walled enclosures dotting the area give good cover for skirmishers. So, with his left covered by the guns mounted on the walls of Almeida and his right 'refused' down by the river bank, Craufurd waited to be attacked when he would give the French a sharp lesson.

Ney, believing Craufurd's position, with the defile to its rear and only a single narrow bridge for retreat, to be faulty, before dawn on 24 July 1810 arrayed his whole corps of 24,000 men in a broad and deep column, fronted by the cavalry brigades of Lamotte (3rd Hussars and 15th Chasseurs) and Gardanne (15th and 25th Dragoons). Then the thirteen battalions of Loison's Division in line of columns; behind them Mermet with eleven battalions; three regiments of Marchand's Division formed the reserve. After taking over an hour deploying for action, the French infantry then came forward rapidly, and the French cavalry, in line of fifteen squadrons, bore down on the much smaller British mounted force, sending them and Ross's advanced guns flying back over the plain. Now came an overwhelming infantry assault when Craufurd's line of three British and two Portuguese battalions were suddenly hit by Loison's thirteen battalions, coming on at the pas-de-charge, their loud cries and shouts rising above the monotonous drum-beating. This first rush was momentarily halted by rolling volleys; in bearskin caps and light-coloured pelisses the French 3rd Hussars braved gunfire from the ramparts of Almeida to sweep across the interval between Craufurd's left and the fortress walls, down onto the flank of the Light Division. Flurried by this sudden charge, the gunners in Almeida fired so wildly as to cause few casualties, and the cavalry swept along the rear of Craufurd's line rolling it up until checked by volleys from the 43rd and the riflemen behind stone walls.

His safest flank turned, Craufurd realised he had to retreat at once; cavalry and guns were ordered to gallop to the bridge, followed by the Cacadores. The remainder of the British infan-

be appreciated that Wellington needed a great deal of transport to supply his artillery. At Quatre Bras and Waterloo his army expended 987,000 musket cartridges, which is not far short of 100 cart-loads.

The organisation of a troop of horse artillery equipped with six 6pdrs is not without interest. These six guns required:- 5 officers, an Assistant Surgeon, and 155 NCOs and men, including the drivers. In addition, the troop required 156 horses; 6 ammunition waggons (1 per gun), each drawn by 6 horses; a forge; a spare gun carriage; and a store waggon, the store waggon being the only 4-horse vehicle in the troop. A field artillery battery, equipped with 6 × 9-drs, called for very similar organisation: 116 horses and 163 officers and men. In this case the gun teams were of 8 horses. It was a complicated business to organise and supply field artillery even in the year 1815.

The knoll on the right side of the road above the bridge over the Coa. On this hillock Beckwith's Light Division Riflemen made a stand to cover the retreat over the bridge. This shot is taken from the corresponding knoll on the other side of the road, on which a similar stand was made

try were to fall back in echelon from the left, defending each enclosure and hillock; the 52nd were to hold fast on the right flank. Marching westward straight upon the British line, smoke and nature of the ground prevented the French seeing Craufurd's enforced change of front more or less to the south, so struck the British line an oblique grazing blow. This was fortunate, because had the French fully hit Craufurd's right simultaneously with closing with his left, they would have overwhelmed the 52nd and reached the bridge before the main body of the retreating force. Coming successively into action, Loison's battalions struck first and hardest against the British left, nearest at the top of the hill; hotly pressed a wing of the 43rd, found themselves trapped within the 10ft (3m) high stone walls of an enclosure and had to throw down the wall to escape.

A fighting retreat is difficult when pressed by an overwhelming foe; and the British companies dared not stand too long in a position for fear of turned flanks cutting off retreat to the bridge, besides having to watch for French cavalry, cantering down the paved road sabreing right and left. Held-up at the sharp turn in the road, the guns and cavalry were further impeded at the bend when an artillery caisson overturned and had to be manually righted; throughout they were harried by French artillery unlimbering on the crest of the ridge to pour shot down on them. They were still blocking the bridge when Craufurd's left was gradually forced down upon them. The situation was eased when Major McLeod rallied four companies of the 43rd on one of the pine-covered knolls above the bridge, while two companies of the 95th positioned themselves on a corresponding hillock on the other side of the track. Holding firm, they allowed Craufurd to deploy guns and Cacadores on the slopes on the far side of the bridge, covering the bridge. Then it was seen that the five companies of the 52nd who had been holding the right wing were still making their way along the river bank. Hoping to cut them off, the French made a supreme effort and dislodged McLeod from his knoll, but he rallied and, aided by everyone within reach, threw the French from the hillock which was held until the 52nd had reached safety. Then dropping down from the knolls, the rearguard ran swiftly across the bridge, allowing French infantry

The Bridge over the River Coa, with the rough track stretching upwards between the two knolls that played such a big part in the combat

to re-occupy the wooded eminences. On the lower slopes of the far bank, commanding the bridge, British infantry strung themselves out behind rocks and walls with Ross's guns above them to sweep the passage.

In the full flush of triumph, just as a wargamer might, Ney decided to force the bridge – but wargamers only lose metal figures whereas the impetuous French General was playing with flesh and blood. French skirmishers came down to the water's edge and from behind rocks, engaged in a lively musketry duel with the riflemen over the river, while guns thundered at each other across the valley. Ney ordered the 66th, a leading regiment of Loison's Division, to carry the bridge; quickly forming, led by Grenadiers the column rushed gallantly forward to be mown down; their bodies reached the top of the bridge's parapets before they fell back. The fiery red-headed Ney now ordered a bataillon d'elite of picked marksmen of the regiments of six Corps to take the bridge, but they only added to the heaps of dead until the bridge was quite blocked. Out of 300 men, 90 were killed and 147 wounded in less than ten minutes, and a third attack by the 66th, delivered with little dash or enthusiasm, was easily beaten back.

A duel of artillery and musketry was exchanged across the valley, until at four o'clock a rainstorm of tropical intensity caused it to cease. Craufurd remained in position until midnight before retiring on Pinhel, having lost 36 killed, 206 wounded and 75 missing in an action he had handled very badly. Ney, had he been wise and contented himself with driving-in the Light Division for small loss, would not have lost 527 men, mostly in mad attempts to rush the bridge.

The rough ground over which Craufurd's Combat on the Coa was fought; the track can clearly be seen and in the left distance is the windmill covering the British left flank

Wargaming the Combat on the Coa

Both commanders lay out their forces as they were at the start of the battle in 1810, preventing distortion of historical

A group of British riflemen at the Coa

aspects through an immediate French blow on the British right by the river, by ruling that Ney must come on from the East and advance directly westwards *until making contact*, when he can change direction after two game-moves.

Craufurd will command the same force as on the day; Ney will have Ferey's Infantry Brigade (of Loison's Division) as they seemed to be the only French infantry who actually got into action, and the two brigades of cavalry that, in the beginning, swept Ross's guns and the British cavalry before them. In full, the respective forces were:–

British – The Light Division composed of 95th Rifles; 43rd Foot; 52nd Foot; 1st and 3rd Portuguese Cacadores; 16th Light Dragoons; 1st Hussars; Ross's Battery Royal Horse Artillery (3 guns).

French – Ferey's Infantry Brigade (of Loison's Division) composed of 66th Ligne (3 battalions); 82nd Ligne (3 battalions); 32nd Legere (2 battalions) and 6 battalion guns; Lamottes Cavalry Brigade formed of 3rd Hussars and 15th Chasseurs, with 1 gun; Gardannes Cavalry Brigade composed of 15th and 25th Dragoons, with 1 gun.

Both commanders can be classified as no more than 'average' (see section 'Classification of Commanders') and there could be grounds for suggesting that Craufurd was, on the day, a 'below-average' leader. However, as it developed into more or less a 'soldier's battle' with each group of British ordering their own movements, if we want to make a game of it, perhaps it is best not to further handicap the already surprised and numerically inferior force!

The objectives of both sides are as they were 'on the day' – the British to survive and get as many as possible over the bridge and out of immediate danger; the French are seeking to prevent this and destroy them. It is a 'strongly recommended' project, as each of three or four reconstructions of this epic contest has produced a quite superb tabletop battle.

A DESIRABLE WARGAMING RESIDENCE

For want of space, villages and towns on wargames tables are usually represented by a church and about three houses grouped around a crossroads or village-green. This is tiresome and can, in part, be avoided by having a wargame in which your entire table-top is filled with the village. Thus, the fighting takes place in the streets and gardens; from house to house. A pleasant effect is obtained by grouping a number of buildings on a baseboard; thus a farm can be made with farmhouse, stables, barns, hayricks, walls, etc, on a board perhaps 3' 0" × 3' 0". This will form a very good focal point for a battle.

However realistic it is made to appear, realism goes out of the window as soon as the wargame commences, when a scaled-down infantry regiment of perhaps 20 figures (representing the 800 odd men who form such a unit) garrison a couple of houses; or an entire army which, in real-life, would total thousands, settle themselves comfortably among the few houses which, on the table, simulates a town or village. Put your mind to any small village of your acquaintance, then imagine how well defended it would be by the 800 men of a single infantry battalion, yet on the wargames table a single unit would never be able to hold such a position. The problem can be easily and realistically solved by building a 'miniature village' –

a colourful collection of houses grouped around a village green on a small baseboard that need only be about 8 inches square – because the buildings are out of scale with the figures, being 1:300 scale houses for 15/25mm soldiers. In this way it is possible to have more than one village on the same table and, as can be seen from the illustrations, it all blends without any incongruity. And it is about the cheapest way of going about the business of buildings, as enough for a complete village can be bought for about a pound-coin, using the 'nets' of small wooden houses sold in those shops catering for very young children (you can always get your mother or wife to actually make the purchase!)

A village is held by a single unit plus one or two guns, but no more; neither guns nor figures are actually placed IN the village, but noted as being there and retained in a marked box by the table. The infantry are considered to be garrisoning houses spread around the village, so that an all-round field of fire is afforded to them; guns must be positioned in a definite site, represented by a small cardboard arrow indicating their field of fire. As infantry are deemed to be defending each house and able to cover, with musketry, every approach to it, they are not moved *within* the village; and guns must remain where placed. Both infantry and guns can move in/out of the village at their normal rate of movement with, if desired, a half-move penalty for 'settling-in'.

Villages can only be taken when stormed by infantry; or by being set on fire by artillery which drives out the defenders; also making the burning village untenable to the attackers! Attackers can fire on defenders who take casualties according to the rules in operation. When artillery fires on a village, defenders in that side or segment of the village under fire are forced to 'keep their heads down' and cannot return fire with their rifles/muskets. *To set a village on fire with artillery* – firer says he intends firing and indicates area of village that is his target; a 'hit' is automatic, then the firer needs to throw a dice, with a one-in-three chance of starting a fire (ie with percentage-dice, he needs 65 or over, with the usual six-sided dice, 5 or 6). Once started, a fire affects that quarter of the village around the point of aim, thus four 'fire-hits' engulf the entire village. In Napoleonic Wars and earlier, only howitzers can start fires, in later wars any guns have the capability. Defenders vacate the burning area the move after the fire starts.

Infantry storming a village – the attackers take musketry fire (and artillery fire if gun is present) –

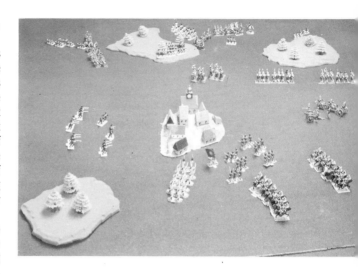

A 'miniature village' set in the centre of a wargames terrain, surrounded by woods and being attacked by columns whose size indicates the reduced scale of the small cluster buildings (Photo by Adrian Jackson)

they make contact with the defenders when they reach the first house on their route to the village. There they remain throughout the ensuing melee, although considered to be within the village for any later rounds of the melee should the defenders do sufficient under the rules to continue resistance. All melees take place with defenders counting as the equal of THREE attacking units ie on a 1:1 basis, but considered to be outnumbered 2:1 when there are FOUR attackers. This can be simply worked by using a pair of Percentage Dice (one red and one black, capable of giving a range from 1-100) – the Attacker throws first and requires a score of 60 to remain in contact; the Defender requires 15. Next move, if the Attacker remains, then he requires 40 and the defender 30. The melee continues as long as the Attackers make their required score, and Defenders hold; it ends when the Attackers are repulsed or the Defenders are forced to withdraw from the village. When the Defender is outnumbered 2:1 (i.e. when there are four attackers then he DOUBLES the score he is required to get), while the Attacker HALVES the score each attacking unit requires; he does the same when attacking on more than one side or face of the village.

Now villages on the wargames table become the bastions they were in real-life when a single battalion well hidden in houses make them the anchorage for the rest of the army.

13
THE MACEDONIAN PHALANX

A Macedonian 'Companion'

The Macedonians, whose kingdom lay to the north of Greece, were a warlike race with Greek stock and traditions but lacking their characteristic culture. Philip, who unified the Macedonians, had spent some years as a hostage in Thebes and had learned much from Epaminondas. On his return, he improved upon the Theban phalanx, increasing its mobility by thinning its ranks to 16 and by lengthening the pikes (sarissa) to 16ft (4.8m), 18ft (5.4m) and even 24ft (7.2m) so that the weapons of men as far back as the fifth rank projected beyond the soldiers of the front rank. To cope with this two-handed weapon, the shield was reduced in size and strapped to the left arm, leaving the hand free. With these exceptions, the arms and equipment of the Macedonian peasant-soldier were not very different from that of the regular Greek hoplite. The usual proportion of cavalry to infantry in most Greek states was about 1 to 12 or 1 to 16, but Philip, believing in the riding habits of his 'farmers', had a higher proportion of about 1 to 7.

Philip was murdered in 336BC and his splendid army came under control of his son Alexander, a man destined whilst still young to become one of the greatest generals the world has known. By breaking the phalanx into smaller units Alexander turned it into a spear-hedged mobile base rather than a juggernaut of moving spear points. But the real striking force of the Macedonians lay in the shock power of their heavy cavalry charging in formation at speed. They were formed of eight squadrons – the Companions on the right wing and Thessalians on the left, both being supported and strengthened by light cavalry and light infantry. Alexander also devised a new class of highly trained foot soldier – the hypaspists – who were a cross between the heavy pikemen of the phalanx and the light-armed peltast. They wore armour and were armed with a shorter spear so that they formed a link between the phalanx and the cavalry. However, there is still doubt as to the actual equipment of the hypaspists. A picked body of these hypaspists constituted the Royal Foot Guards (the Agema).

In battle the Macedonian phalanxes advanced in echelon so that the right division struck the enemy first and 'fixed' him. Then the Companions, often under Alexander himself, attacked the enemy's left supported by the hypaspists. If the enemy attempted to outflank the left of the Macedonian phalanx then the Thessalian cavalry, together with light cavalry and light

infantry, opposed them whilst a similar screen guarded the right of the Companions and moved up to engage the enemy left while the heavy cavalry attack succeeded. It was a system that required mutual support of all arms arising from a sense of confidence in each other and in their leader.

There was only one unit formed of lancers in the Macedonian army, cavalry armed with the sarissa, but the rest would normally have the 'standard' short spear, sword, and javelins in some cases. It was the training and discipline, combined with Alexander's own leadership, and timing, that made the Macedonian cavalry so formidable – even then, it is doubtful if they would be able to break steady hoplite infantry in a frontal charge.

The Graeco-Macedonian armies reached their peak under Alexander, who had the ability to arouse great loyalty amongst his followers, together with a supreme skill in freely manoeuvring his well trained army in small units – one of the great developments of the art of warfare to come out of this period.

Persian 'Immortals' enemy to the Macedonians

Double-dealing on the wargames table

Treachery has been with mankind since the beginnings of time and while it may be controversial to assert that it was more common during the Ancient and Medieval periods, it would seem that this factor can be utilised to give a new facet to wargaming. The likelihood of treachery can be built into a campaign narrative or, before a battle commences, some indication can be given that an army, part of an army or even an individual unit is suspected of treacherous motives and may well defect. In this case a commander may be given a 'Treachery Option' that he can exercise at any stage of the battle. To counter this some sort of a factor may be devised that enables the commander of an army to attempt to win back the loyalty of a treacherous unit. The effects of treachery could be that a treacherous unit or part of an army leaves the battle or it may defect to the enemy. In the event of a treacherous unit becoming panic-stricken so that it is routed, then it is beyond being rallied. As with most situations in wargaming, the easiest solution is to throw a die but this relies far too much on luck and the following 'Treachery Formula' is suggested.

First it has to be decided whether it is an entire army, a wing or a separate unit that is likely to display treachery. Having reached that conclusion then certain decisions have to be made:
continued

Alexander's army on the march during the invasion of Asia Minor, when they reached the valley of the Ganges in India

Style of Fighting

Philip of Macedon is said to have created the first scientifically organised military force of history when he used heavy missile engines in co-ordination with the field operations of infantry and cavalry – it was the forerunner of field artillery.

Alexander used an oblique order of battle, echeloned back from the right flank cavalry spearhead or with the hypaspists on the right flank of the phalanx to provide a flexible hinge between the fast-moving cavalry and the relatively slow hoplites. Flanks and rear were protected by light infantry, peltasts drawn up in a line approximately eight men deep behind the phalanx and psiloi (servants and foragers for the heavy infantry) formed a skirmish line in front; they were armed with bows, javelins, darts and slings. Besides being a base of manoeuvre for the shock

action of his cavalry, the heavy infantry phalanx was a highly mobile formation, capable of attacking at the run whilst retaining formation so as to hit the disordered enemy ranks before they could recover from the impact of the early cavalry attack.

Alexander the Great was a military genius, capable of adjusting his tactics and tactical formations to suit prevailing conditions. His tactical skill in mountain warfare and against irregular forces was of the highest order and he must be considered to be one of the three greatest military leaders in history. His father Philip of Macedon was also an extremely good general and could be classed as an 'Above-average' commander.

Warfare is littered with examples of seemingly unconquerable strategical and tactical ploys that have come unstuck – and the feared Macedonian phalanx was no exception. Coming up against the relatively unknown strength of Rome in her early years, the Macedonians were given the shock of their lives at the battle of Cynoscephalae in 197BC. Fought five years after the great battle of Zama and the first occasion on which the flexible Roman legion and the Macedonian phalanx met in the open, the affair at Cynoscephalae was the Jena of Macedon, revealing to the incredulous Greeks that the phalanx had met its master, that the descendants of Alexander's soldiers had given way at the first shock to the 'unknown quantity' of the Roman army. This supreme test of the merits of the rival formations occurred during the Second Macedonian War (200–196BC), fought in Greece and Asia. It was a conflict in which the successors of Alexander of Macedon took on the increasingly powerful Romans, who were allied to most of the Greek cities. Arriving at Heracles in April 197BC, Titus Quinctius Flaminius linked up with the Aetolian army of 6,000 foot and 400 horse under Phaeneus, and with Amynander and 1,200 Athamanians, bringing his army up to about 26,000 men, including 2,400 cavalry. Flaminius led them into Thessaly as Philip V of Macedon moved towards him from Larissa at the head of an army of some 23,500 foot (including 18,000 Macedonians) and 2,000 horsemen.

There was some desultory fighting near Pherae before both armies turned west seeking better ground, and lost touch with each other, although they were marching on parallel lines – Philip to the north and Flaminius to the south of the Cynoscephalae hills (Kardagh). In these hills near Scotussa, on a misty morning, covering detachments of both armies clashed in a small action that grew larger as reinforcements came up, until both sides were heavily engaged. The Romans were pressed back, despite some bold charges by the Aetolian horse, as Flaminius deployed his army facing the hills, before advancing his left to meet the enemy. Encouraged, the Macedonians moved forward to occupy the southern slopes of Kardagh, although the unexpected general engagement was not entirely to Philip's liking, as the Macedonian left wing had yet to come into position and, more important, the ground over which they were to fight was broken and unfavourable to the phalanx.

1. Is the commander of that force for or against treachery?
2. Are the troops forming the force for or against treachery?
3. Is the bribe offered to the unit high enough to persuade them to be treacherous?

These are basically yes-or-no choices and can be decided by the throw of a die, thus a die scoring 1, 2 or 3 means that a commander is for treachery whilst a 4, 5 or 6 score means that he is against it. A score of 1, 2 or 3 means that his troops are for treachery, 4, 5 or 6 they are against it. A score of 1, 2 or 3 means that the bribe is high enough to make treachery attractive whilst a score of 4, 5 or 6 means that the bribe is not high enough.

Each 'yes' (for treachery) gives a -1 factor to the unit, each 'no' (against treachery) gives a $+1$ factor to the unit.

Now must be considered other relevant factors:

Does the unit have any national affinities to the people to whom it intends to defect? Yes -1; No $+1$.

Is its position on the battlefield conducive to defecting?
  (a) It is on a flank and can get away -1.
  (b) It is flanked on both sides by other units and can only go forward $+1$.
  (c) It is not a forward position and has other bodies of troops around it, making defection difficult $+2$.
(This situation might well be altered if any of the units around the one under consideration have already defected or indicated that they are likely to be treacherous.)

continued

153

Is their side losing? Yes −1; No +1.

Having totalled up the pluses and minuses and arrived at a figure, throw two dice, one coloured and one white – take away the score on the coloured die from the score on the white die and add or substract that figure to the total already arrived at.

If resulting total after all factors have been considered and defensive dice added or substracted equals 0 or a minus figure then the unit is treacherous and defects at charge-speed rate of movement and in any direction it wishes.

If the final score is plus one or more then the unit is not treacherous and carries on as before but if this figure only totals 1 or 2 then the unit will test for treachery again on the following move.

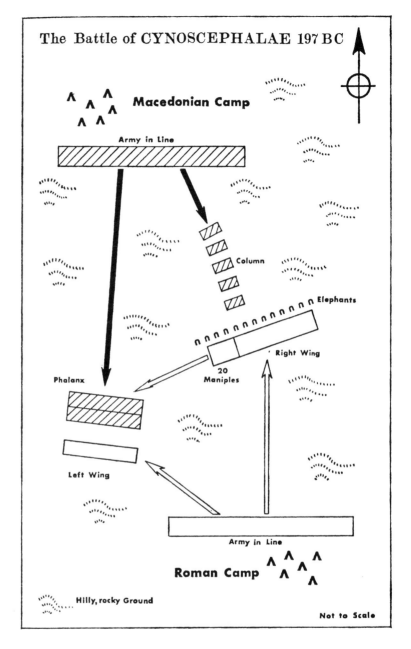

The Battle of CYNOSCEPHALAE 197 BC

The battle fell into three separate and successive actions. First, on the west, Philip descended from the hills with the right half of the phalanx to drive back in great disorder the Roman left, personally commanded by Flaminius. At this reverse the Roman commander rode across to his hitherto inactive right wing and ordered the legions and their allies, preceded by some elephants, to fall upon the left half of the phalanx, at that moment deploying from march column. Disordered by the broken ground and fearful of the elephants, the Macedonian wing in this second part of the battle could not cope with the attack of the legions and the Aetolian infantry, and was broken

154

and routed. The Roman right-wing formations surged forward in pursuit, and a Tribune, acting on his own initiative, detached 20 maniples, totalling 2,000 men formed of the principes and triarii, and led them to the left. Outflanking the victorious Macedonian right-wing phalanx, the maniples in this third and final period were able to attack the formation from the rear, breaking it and driving it from the field in confusion.

Realising that all was lost, Philip rallied the survivors and left the field, having lost about 13,000 men against Roman casualties of a few hundred.

Reconstructing the Battle

The battle can either begin with small parties of light troops and cavalry skirmishing in the centre of the table as they endeavour to win superior positions for their advancing armies, or with Flaminius deploying his army and advancing his left as Philip's right-wing phalanx moves forward to attack him. This first phase of the battle is followed by the Roman and allied infantry on the right, supported by elephants, catching the Macedonian phalanx before it has formed. The third phase begins with the Tribune leading his 20 maniples against the Macedonian right-wing phalanx, which has already defeated the facing Roman formation. Too much time must not be allowed to elapse here, as no self-respecting wargamer is going to allow a victorious formation to stand and gloat while its comrades are being beaten.

Commanders' Classification

Neither Philip nor Flaminius showed up particularly well: the former accepted battle knowing the terrain to be unsuitable to his formations; and the latter was beaten on the left, though he deserves credit for sending his right wing into action after this defeat. Both may be rated as Average, but the unknown Tribune who led 20 maniples into the rear of the Macedonian phalanx should be classified as Above Average.

Quality of Men and Style of Fighting

The luck of the Romans at Cynoscephalae in encountering a hastily trained force operating on unsuitable terrain encouraged the legions to find their feet at an early formative period. There is no gainsaying, however, that the relatively fast and mobile legion possessed a tactical flexibility denied to the solid mass of the now slow and unwieldy phalanx, particularly in this battle when the latter was formed of ill-trained and inexperienced troops operating on rocky ground. Like the British square, the strongest feature of the phalanx was its cohesion; if thrown into disorder by the terrain or caught in the process of forming up, it was lost. In favourable conditions it pushed forward like an irresistible militant hedgehog to carry all before it.

The superiority of the Roman military system was emphasised a generation later at Pydna, when the legion again beat the phalanx, this time on level ground well suited to phalangial

---

**Favoured Formations for Wargaming –
The New Model Army**
When Parliament brought into existence in 1645 the New Model Army, England at last had a regular and, later a standing army. Owing their appointments to their fighting records, its generals were Sir Thomas Fairfax, Commander-in-Chief; Oliver Cromwell, Lieutenant-General in command of cavalry and Philip Skippon commanded the infantry. Fairfax, 33 years old, was undoubtedly the best man to be captain-general. He was a good soldier and a gallant man whom everyone was pleased to serve under. Philip Skippon was a veteran of the Low Countries campaigns and much respected.

The regiments received new clothing and soon, for the first time in English history, the army were dressed in the familiar scarlet; facings of the Colonel's colours distinguished the various regiments with the senior corps – the Captain-General's own – wearing blue. Formed from the Eastern Association cavalry trained by Cromwell, the horsed arm were first class volunteers, all well mounted and armed. The infantry were not of the same standard, about half being pressed men and their ranks included many ex-Royalist prisoners-of-war. The artillery was, for its day, remarkably efficient and is said to have consisted of 16 demi-culverins, 10 sakers, 16 drakes and 15 smaller field pieces plus mortars and siege-guns. Light field pieces were usually attached to infantry regiments. The paper strength of the New Model Army was 22,000 men divided into 11 regiments of Horse in 6 troops each of 100 men with captain, lieutenant, cornet and quarter-master and 3 trumpeters to a troop; each regiment had a colonel and
continued

major as field officers. A Dragoon regiment of 1,000 men was divided into 10 companies each of 100 men with 4 officers and 2 drummers. Dragoons were armed with the musket, formed in 10 ranks of 10 men abreast, and obeyed the drum; in action 9 men fought and the tenth man held their horses. The 12 Regiments of Foot each had 1,200 men in 10 companies divided equally into pikemen and musketeers; the Colonel's company of 200 men, the Lieutenant-Colonel's 160 and the Major's 140; the other 7 companies each with 100 men. There was a captain, lieutenant, ensign and 2 drummers to each company. All regiments – horse, dragoon and foot – were known by the names of their colonels. The infantry's normal formation was not less than eight ranks deep; New Model cavalry fought in two ranks trusting to speed to overcome weight; the 'rank' was substituted as the tactical unit of the troop in place of the 'file'. They won their first battle at Naseby where, although Rupert overthrew their left wing and Skippon was wounded early on, Fairfax's regiment stood firm and Cromwell's cavalry were irresistible. The New Model Army continued to perform well in the Civil War and later against the Scots at Preston and Dunbar; also in 1658 at Dunkirk Dunes during the Spanish-French Wars.

tactics. This defeat of Perseus (Philip's son) has been accepted by many military writers as conclusive of the merits of the two formations. Yet the time was past for the trial to be a fair one, as the phalanx was no longer at its best, partly because the length of the sarissa had been increased to 21ft (6.3m), so limiting the manoeuvrability of the phalanx, as the men had to bunch tightly together for mutual support, grasping their huge spears with both hands. Even more incapable than before of changing front, the formation immediately broke up if assaulted in flank or rear. The contemporary practice of relying on unaided infantry to protect the flanks was not as successful as Alexander's use of cavalry for this purpose. The phalanx could still perform wonders on level ground against an adversary who awaited attack, but when opposed to the mobile and flexible legion, resembled a bull assailed by nimble matadors. This tactical flexibility had been given to the legion by Scipio Africanus, and the troops employed by Flaminius were largely Scipio's veterans from Spain and Africa where doubtless the Tribune who turned the battle had learned his tactical lessons. It was on the foundations laid by a more gifted commander, therefore, that Flaminius gained his victory at Cynoscephalae.

It can be claimed that the maniples led forward by the un-named Tribune played the biggest part in the victory. The 20 maniples (a maniple was composed of 2 centuries each of 60 to 80 men, with the triarii maniple consisting of 1 century only) were formed of principes or veterans, the experienced backbone of the army (armed with 7ft (2.1m) javelins and a broad-bladed short sword or gladius), and the older triarii. The javelins were thrown at the enemy just before contact, and then the principes closed with the sword in a tactical concept comparable to a bayonet attack preceded by rifle fire. The triarii were the oldest group of legionaries who, forming every third line of infantry, brought steadiness and experience to the formation. They were armed with a 12ft (3.6m) pike and the gladius.

The Aetolian allies of the Romans are frequently mentioned in accounts of the battle and must have been first-class troops.

The number of elephants is not known, but undoubtedly they played a big part in scattering the forming Macedonian phalanx. Elephants possessed both enormous potentialities and great limitations. Horses often refused to face them, but disciplined formations could either turn them back or allow them to run through intervals in their ranks.

Morale

Both forces could start the battle with the same standard of morale, or the Macedonians, a relatively scratch force, might start with a lower morale than the Romans, but the morale of each in turn must be affected by the fluctuating progress of the battle. Morale rules could be slanted to allow the Macedonians to fight well until the moments of crisis, when they disintegrate rapidly. These moments are when the left wing encounters the elephants, and when the victorious right wing is hit in the rear by the maniples, both circumstances requiring strong and immediate morale-reaction tests.

Terrain

All important in this engagement, the terrain was sharply undulating and extremely rocky, as might be expected on the slopes of a range of hills. It is probably best simulated on the wargames table by numerous low piles of books, pieces of wood or plastic ceiling tiles covered by a blanket or cloth. The surface should be liberally spread with pebbles and pieces of stone, scrub and small stunted trees to simulate the uneven ground that helped to destroy the phalanx. While your bemused opponent is still studying this tricky terrain, quickly assume the mantle of the Roman commander!

Military Possibilities

Of course, the prime possibility would be for Philip not to accept battle on unsuitable terrain, but that would mean no wargame. Other possibilities are:

1. The victorious Macedonian right phalanx has the time or experience to exploit its success.
2. The Macedonian left-hand phalanx forms more quickly, and is caught in relative disorder by the legions and elephants.
3. Tribune is not so alert, and does not lead his 20 maniples to attack the Macedonian right phalanx.

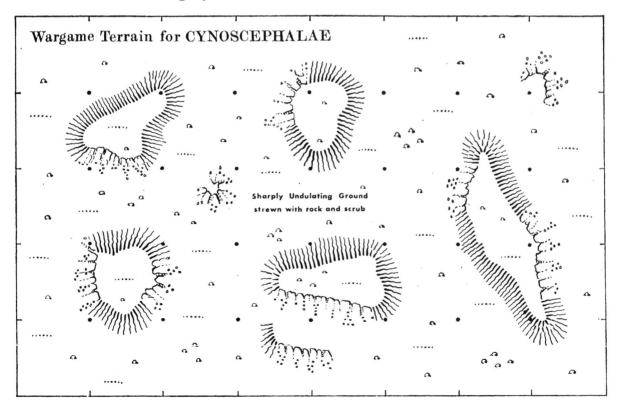

Wargame Terrain for CYNOSCEPHALAE

Sharply Undulating Ground strewn with rock and scrub

PRACTICAL NOTES ON PONDEROUS PACHYDERMS

Initially war-elephants were used by Eastern nations, coming forcibly to Western notice when Alexander came up against them at the Hydaspes in his 326 BC invasion of India, where both limitations and potentialities of the beasts were amply demonstrated. When Alexander's cavalry-horses refused to face the elephants, it was his disciplined phalanx, surmounting their fear and surprise, who eventually turned the ponderous pachyderms back in panic-stricken flight. Twenty years later Seleucus' ceded territory to Chandragupta in exchange for 500 elephants, which brought him victory at Ipsus. Subsequently their use spread to Greece and Carthage when such farseeing commanders as Hannibal used them. At Beneventum and Heraclea elephants were most successful against troops unacquainted with them, particularly if those troops were not particularly well disciplined. Trained, professional troops could handle them – the Romans opened ranks and allowed the animals to pass through. Frightened by

the noise, goaded by arrows and spears, its driver killed or wounded, a stampeding elephant was as great a menace to its own side as to the enemy. To prevent this, the elephant's mahout (driver) was armed with a steel spike to hammer into the beast's brain when a stampede seemed likely.

At Heraclea in 280 BC, against the Romans, Pyrrhus King of Epirus, brought up 20 war-elephants at a time when the battle was turning in Roman favour. Never having seen such monstrous animals before, the Romans were thrown into confusion so that their fleeing cavalry pursued by the elephants broke the ranks of the legionaries and enabled the Greeks to gain the upper hand. However, the losses were so great on both sides that it gave birth to the term 'a Pyrrhic victory'. Although Pyrrhus again defeated the Romans in 279 BC he was heavily defeated at Beneventum in 275 BC when the Romans actually captured four of his war elephants.

Elephants rarely 'crashed' through infantry at high speed, they were more inclined to amble forward at infantry-charging speed. They tended to force units out of line rather than trample them, although this occurred once a unit broke and ran. They were also vulnerable to archers, who could pick off the mahout (driver) and the bowmen in the elephant's howdah.

Ancient Indian elephants – 25mm Essex figures. Buildings Hales Models (Courtesy of Duncan Macfarlane, Wargames Illustrated)

Cavalry rarely attacked elephants because horses did not like going near the big animals; a line of elephants was sometimes used as a barrier against cavalry. Horses cannot discriminate in their dislike of elephants so that cavalry can only work with 'friendly' elephants if they have become accustomed to the presence of these 'ponderous pachyderms'.

Experienced in warfare against elephants, Indian troops found fire-arrows most effective; at Gaza, Ptolemy used what was perhaps the very first anti-tank minefield – iron spikes, chained and anchored to the ground, to rip the tender feet of the elephants!

When wargaming, elephants are probably the most tricky facet of ancient warfare, either to fight with or fight against. They can be devastating to the enemy, but they can also devastate one's own side! It requires judgement to place them and when to let them loose, and an equal judgement if opposing them.

The fatal mistake when fighting against elephants is to be too fearful of them, and to let them dominate a battle-plan. It is no good infantry packing together into a mass 5 or 6 ranks deep surrendering the initiative to the enemy – because he will simply use his elephants to keep your infantry pinned in that imobile mass while he uses the remainder of his army to wipe out the rest of yours. Then he can then attack your infantry from all sides and destroy them also. Don't over-rate elephants!

The best weapon against elephants is massed missile fire, doubly dangerous to the owner of the elephants as a wounded elephant is liable to stampede back through its own army. True, elephants are not easy to knock-out by arrow-fire, but if archers are massed against them, a percentage of hits must be damaging ones. The most deadly enemy of the elephant is the horse-archer, particularly if his horse is inured to elephants; they can ride up to close range, pour in deadly volleys, and be away again before the ponderous elephant can react. However, the experienced elephant commander will not readily expose his valuable beasts to enemy horse archers. While every endeavour should be made to get at the elephants, do not lose sight of the fact that the mere presence of horse-archers is neutralising the elephants and forcing waste of infantry in screening them. Use this advantage to gain a local supremacy elsewhere.

Never, save as a last resort, engage elephants with heavy cavalry; even if they are inured to elephants, their chances of achieving anything

War elephants (by Miniature Figurines, Dick Higgs and John Chapman collection)

against them are strictly limited, and a whole unit of cavalry can be lost without killing a single elephant. Attack elephants with light infantry armed with spears, who stand a good chance of doing damage without excessive loss to themselves. Never allow pre-occupation with enemy elephants to distract attention from other parts of the field.

From the foregoing it will be seen that the path of the general using elephants is not free from pitfalls; he will be well advised to adapt use of them to circumstances. If the enemy is well equipped with missile-troops, the role of the elephants may be that of a constant threat, tying down a proportion of the enemy army while the real blow is delivered elsewhere. Above all, they must not be stationed in advance of their own main line, where they can do vast damage if stampeded. If the enemy is short of missile strength, then the elephants can revert to their original role of tanks, being launched straight at the enemy to achieve a breakthrough.

Do not keep elephants concentrated in one place; spread them along the front. A single elephant can often do just as much damage as a group of five and one elephant charging into an infantry regiment is as likely to break it as are a number; by spreading your attack, the law of averages brings success somewhere. On the other hand, the whole group of elephants can guard one wing, where they can both pin down and damage the enemy, with no danger to their own troops.

Do not discount the value of a sudden surprise

20mm Egyptian elephant (De Gre collection)

stroke with elephants at the outset of the battle, which may well catch the enemy on the wrong foot and achieve victory almost before the action has really begun. This is not a tactic to be used too often, being in the nature of a gambler's throw, losing elephants without achieving any real success in return.

Here are some well-tried rules controlling the use of elephants, an interesting adjunct to tabletop battlefields.

1. Elephants attacking infantry in three ranks. Dice throw of 1 makes elephant swerve. If infantry in more than three ranks, dice throw of 1, 2 or 3 makes them swerve. Second dice throw decides direction of swerve – 1 or 2 right, 3 or 5 left and 5 or 6 straight back. Elephant is then deemed to have stampeded.

2. Stampeding elephant. Any individuals in path of stampeding elephant are killed, any unit taken thus in flank or rear will lose a quarter of its strength and must throw for its morale. If a stampeding elephant runs frontally into an organised body of heavy infantry or cavalry, dice throw to see where it goes. 1 or 2 swerves right and continues in that direction moving further each game move, 3 or 4 swerves left and similarly

continues, 5 or 6 plunges straight through unit, destroying one file, and unit must throw for morale.

3. Elephant attacking home on heavy infantry. Provided the elephant does not swerve, he will charge home on the ranks of infantry who must throw a dice to see if they withstand the charge. Infantry in more than 3 ranks need 3, 4, 5 or 6 to fight, infantry in 3 ranks need 4, 5 or 6 to fight, infantry in less than 3 ranks need 5 or 6. If the infantry do not stand but break each man must have a dice thrown for him – 4, 5 or 6 to escape, 1, 2 or 3 trampled upon and out of battle. If infantry stand, throw one dice for every five men; if a 5 or 6 is thrown on any one dice throw again and second 5 or 6 in succession hamstrings elephant, putting it out of battle. One dice is thrown for each elephant and total scored represents men killed, thus a 5 on dice means five men killed.

4. Elephants against light infantry or heavy infantry in less than three ranks. Infantry need 5 or 6 to stand and fight, melee then carried out as above.

5. Elephants against cavalry. If cavalry belong to army that does not habitually contain elephants, so that the horses are unused to them, then the cavalry must throw 4, 5 or 6 to stand. If horses are used to elephants then they do not need to throw dice but stand automatically. If cavalry stand, 4 of them fight each elephant with straight dice throw, highest score wins. If cavalry beaten they are killed, if elephant beaten throw another dice – 4, 5 or 6 elephant unhurt, 2 or 3 soldiers on elephant killed, 1 elephant killed. If cavalry break, dice throw for every man – 4, 5 or 6 saves; 1, 2 or 3 killed.

6. Elephant against elephant. Straight dice throw, highest score wins and loser is killed.

7. Light troops remaining within 12 inches of elephant. Without 3 ranks of heavy troops between them must throw for morale at end of game move and before they can fire. They are shaken by 1, 2 or 3 and must again throw in the same way as ordinary infantry.

8. Elephants hit by missiles. Elephants can be hit by siege engines at normal range, but no further than 12-inch range by hand missiles.

Hits by siege engines, throw dice: 1 or 2 elephant and riders killed. 3 or 4 elephant wounded, remains stationary, riders unhurt. 5 or 6 elephant unhurt.

Hits by hand missiles, throw dice: 1 stampedes. 2 killed but men unhurt. 3 wounded, remains stationary. 4, 5 or 6 unhurt.

Well-handled, elephants can be a very valuable addition to any army – if it is one historically known to have employed them – but they are a weapon whose use requires thought and experience.

14

THE INCOMPARABLE ENGLISH ARCHER – Forerunner of the British Infantryman

At some time in their history nearly every race on earth had used the bow and arrow, but nowhere did they reach the pitch and skill and perfection as in England during the 14th and 15th centuries. In that period the English bowman dominated the wars of Europe as no comparable force has ever done since. The mere sight of them was enough to strike the fear of God into an enemy who, if he did not retreat or keep his distance, was almost certainly slaughtered. Somehow the enemy never learned this. French, Irish and Scots – each of them lost thousands of their youth and nobility to the laconic English bowman. All the archers in Edward III's army were Englishmen. They were strong, muscular men; sinewy, brown, clear-eyed and hard-visaged – middle-sized or tall men of big robust build, with arching chests and extraordinary breadth of shoulder. Their profession was proclaimed by the yew or hazel stave slung over their shoulder, plain and serviceable with the older men but gaudily painted and carved at either end when belonging to younger archers. Steel caps, mail brigandines, white surcoats with the red Lion of St George, and sword or battleaxe swinging from their belts completed the equipment. In some cases the murderous maule or five-foot mallet was hung across the bow-stave, being fastened to their leathern shoulder-belt by a hook in the centre of the handle. When they went to war, spare bow-staves were taken, plus three spare cords allowed for each bow and a great store of arrow-heads.

English archers carried into the field a sheaf of twenty-four arrows, buckled within their girdles. A portion of them, about six or eight of them, were longer, lighter and winged with narrower feathers than the rest.

An archer was often called upon to shoot straight and fast; but often he had to deal with an enemy hiding behind a wall or an arbalestier with his mantlet (a wooden shield) raised – the only way in which such protected men could be hit was to fire in such a manner that the shaft fell upon them straight from the clouds. Even as early as Richard I's siege of Messina, the archers drove the Sicilians from the walls in this manner – 'for no man could look out of doors but he would have an arrow in his eye before he could shut it'.

It is evident from the fact that they wore so little defensive armour that the archers were designed to be light infantry, swift and mobile, skilful and deadly with their weapons. The English

A 54mm model of an English 15th Century Archer

161

*A longbowman 25mm figure
(Colin Dix collection)*

archer was reputed to be able to draw and discharge his bow twelve times in a single minute, at a range of 250yd (225m), and if he once missed his man in those twelve shots he was but lightly esteemed.

For a weary and sick army of less than 6,000 men to defeat over 25,000 French at Agincourt must indicate that the archers could notch with a shaft every crevice and joint of a man-at-arms' harness, from the clasp of his bascinet to the hinge of his greave.

English archers tried to avoid fighting with the sun in front of them, considering the dazzling splendour of a summer's day to be very unfavourable to shooting. At Crecy, when the sudden gleam of sunshine after the rain burst forth behind the English, its beams, besides dazzling the eyes of the enemy, flashed upon their polished shields and corselets with a lustre so brilliant that the archers discharged their first flight of arrows with more than usual certainty of aim.

In addition to being incomparable with his missile weapon, the English archer would frequently discard his bow and fight on foot with sword, axe or maule. When a knight was seated on a horse it was almost impossible to get any power into a swing with a sword, so that he had to stand up to deliver his blow. Standing in the stirrups, he left exposed the one unprotected part in his whole armoured body – his seat. This was the target of the nimble archers and they seldom missed with their keen swords as they dodged on light feet in and out of the horse and foot melee. Well could Sir John Fortescue say: 'The might of the realme of England standyth upon archers.'

The English army had lifted itself from the dragging chains of the feudal system to become a paid, professional short-service army in which the mounted noble and the yeoman archer served overseas under contract to the King. It was a highly trained and disciplined mercenary army; a soldier drawing regular pay for his services is more amenable to discipline than the man dependent on looting and plunder. Edward's army was the most powerful and highly trained force of its day.

In the matter of arms and armament the soldiers of both countries were not dissimilar. Both sides had men-at-arms (knights were men-at-arms but men-at-arms were not necessarily knights) armed similarly with sword, lance, dagger and sometimes battle-mace, helm, shield and spurs completing the equipage. A knight had three armed attendants, who might be pages, to clean and polish his armour, help him in and out of it, hold his horse and assist him to mount; they also groomed the horse; then he had two mounted archers and one swordsman, the whole constituting a 'Lance'. He also had three or four horses, including two heavy chargers (destriers). From the end of the 13th century the horses themselves wore defensive armour. A chanfron protected its head whilst the neck was covered with a crinet with mail attachment and the front of the horse's body was protected by the peytral, its sides by the flanchards and its rear by the crupper. A strong horse had no difficulty in carrying this defensive covering which in the later stages of its development only weighed just over 70lb (30kg), including saddle and mail.

Men-at-arms were covered in armour from top to toe (cap a pie). The increase in plate-armour reduced the mobility of the men-at-arms as it reduced the effectiveness of the arrow. There is plenty of evidence in the chronicles of the French Wars that if men-at-arms, covered completely in plate, advanced against English bowmen without their too-vulnerable horses then they would stand at least some chance of coming to handstrokes. When a body of fully armoured men-at-arms plodded with bent heads into the storm of arrows, however powerfully the shafts struck the hard, smooth, curved surfaces of the armour, they would glance off unless they found lodgement where plate overlapped plate. The effectiveness of plate-armour gradually made the shield unnecessary – it was ineffective against cannon-balls, anyway. Armour improved slowly from about the middle of the 13th century when mail was worn, with a flat-topped

A medieval impression of the Battle of Agincourt 1415

English Archers – 100 Years War

**Wargaming the
'Pike and Shot' Period**
In the early fifteenth century
artillery and the hand-held
firearm were rapidly superseding
the bow and edged weapons, and
for the next 250 years 'pike-and-
shot' prevailed. Gunpowder so
dominated the battlefield that
armour, except for the helmet
and breastplate worn by heavy
cavalry and pikemen, was
discarded. It was an era that had
many positive forward steps in
the art of warfare and is
particularly attractive for
simulation in miniature because
of its colour and brilliant
commanders; it is packed with
tactical innovations, and many of
its battles were fought in
compact areas with small
numbers of men arriving on the
battlefield because of the
dificulties of travelling on the few
bad roads that were available,
and because of an often less than
perfect intelligence system, the
need to protect communications,
and the chronic financial
troubles that made it hard even
to pay national troops, let alone
hire mercenaries. The days of the
massed levy had yet to come,
and cash was often the
controlling factor over the size of
an army.

Many of the 15th-, 16th- and
early 17th-century battles were
worthy of reproduction being
coloured by highly competent
soldiers – Henry of Navarre,
Gustavus Adolphus, the young
Enghien (later to become the
great Conde), Montrose, Maurice
of Orange, and Turenne.

In practically every battle
reconstruction the actual
numbers of men can be
conveniently ignored, provided
that there is a body of troops on
the wargames table to form
continued

barrel helm; then from about 1280 it was reinforced with plate
and the helm was 'sugar-loaf'; from 1300 there was further plate
reinforcement, and a visored helm (the great bascinet), and in
the 15th century complete plate-armour was common – this was
undoubtedly the finest period of armour. French archers were
armed with the crossbow, more powerful than the longbow but
four arrows could be fired in the time it took to discharge one
bolt. Usually used by Genoese mercenaries, it was more accurate
and had a shorter range.

Up to now the longbow had been employed principally
in defensive warfare against an enemy numerically inferior in
cavalry. But when Edward III led his invading force into France
the conditions of war were entirely changed in that they were
up against a country invariably superior in the numbers of
their horsemen. The yeoman with his longbow was soon to
find that the charging squadron presented an even better mark
for his shaft than the stationary mass of infantry formed by the
Scots schiltron. At the beginning of the Hundred Years' War, in
the early 1340s, the Continental world had not yet learned that
it was almost hopeless for cavalry to try to force, in a frontal
attack, a position defended by men-at-arms supported on their
flank by archers. The Battle of Halidon Hill formed the prototype
for Morlaix, the first pitched land battle of the Hundred Years
War and all the other great battles of the war – except the last.
At Morlaix, the Earl of Northampton was victorious, against odds
of 4–1, using Halidon Hill-style tactics that were to be followed
by the English throughout the war – the men-at-arms fighting
dismounted, the trench in front forming an obstacle (a marsh
at Halidon), the defensive position on a ridge, and the skilful
use of archer's fire-power in co-operation with supporting heavy
troops.

Almost for the entire period of the Hundred Years' War, when
the French attacked on horse, they packed their men-at-arms

into a close and solid mass with ranks and files closed up as tightly as possible to maintain a compact array. Even if the English archer was not always able to completely prevent the French attack striking home, he was able to decimate its ranks so that it was weak and disordered when it reached the English position. A man-at-arms was not a headlong galloping cavalier, his attack could not be very rapid unless it was made in disorder; it was shock-action, but shock of a ponderous column moving at a moderate rate. Moreover, if the arrows did not penetrate the armour their effects were such as though they did, for the presence of archers in the field eventually compelled the French to advance on foot. Though plate-armour is not much heavier than mail and is most flexibly jointed, it is not meant for marching in. The necessity of having to trudge a mile or more, often uphill or over ploughed land or through long grass and scrub (as at Mauron in 1352 and at Poitiers in 1356), and to fight at the end of it, was almost as devastating to the French men-at-arms as having his horse shot from under him. More often than not, he died in either case. It is most marked that in all the English victories during the Hundred Years' War it was always the French who attacked and trudged up the hills in their armour. The English quietly stood about, waiting in their strong, carefully chosen defensive positions, perfectly fresh for combat when the exhausted Frenchmen came to grips with them.

Along with the Swiss pikemen, the best professional fighting man of his day, the English yeoman and his longbow were the most significant single factor that changed all the old traditions and concepts of medieval fighting and warfare. It reduced war to two simple elements, one or both of which have to be employed to defeat an enemy – he must be overthrown either by shock, or by missile fire, or by both in combination. The shock method means that success is achieved when one side betters another, often through superior numbers, in a hand-to-hand struggle. This method is materially affected by the superiority of arms or the greater strength and skill with which they are wielded. The missile method means that the day is won by one side keeping up such a constant and deadly rain of missiles that the enemy are destroyed or driven back before they can come to close quarters – in this manner a smaller force can defeat a larger one. Both methods are capable of numerous variations and techniques and combinations of various arms and tactics.

Archery was simply a primitive form of artillery, playing the same part then as now by softening up the enemy to allow the infantry to get to grips under the most advantageous conditions.

The early 14th century saw the evolution of a coherent military system which used in a single tactical scheme the distinctive power of archery, the defensive solidity of dismounted men-at-arms and, when necessary, the offensive power of mounted troops. Edward I realised the virtues of archery in attack to break up a defensive infantry formation and its power in defence when based on a formation of dismounted knights and

corresponding tactical formation taking part in the battle. Thus, at Cropredy Bridge, Waller put into the field 3 bodies of cavalry totalling about 2,500 troopers and about the same number of infantry, a force that, with any conventional scaling down, requires a fair number of figures. But they can be represented by 3 bodies of cavalry of any number to simulate the forces of Middleton, Hazlerig and Vandruske, and 4 separate bodies of foot soldiers to represent the 18 companies of Parliamentary infantry that took part in the early stages of the battle. As these infantry were chased off the field, the same figures can be used to represent the Kentish and the Tower Hamlet Regiments, so successful in the latter part of the engagement. In those battles where only part of an army was engaged, or its separate parts engaged at varying times, the same group of figures can be used more than once, so drastically reducing the numbers required for the battle.

A Man-at-Arms – Early 100 Years War period 1336–1453

men-at-arms. He realised a cavalry attack could be weakened, almost to annihilation, by volleys of arrows. Such knowledge, at a time when cavalry held absolute supremacy in war, was of unfathomable value and laid the foundations of England's military power.

Directed by brilliant, brave and far-seeing captains, the English army did not lose a major battle between Morlaix in 1342 and Patay in 1429.

The successors of the English archer fought with Marlborough at Blenheim, Wolfe at Quebec, Clive at Plassey and with Wellington in the Peninsula. Their bones also salted the Sudanese sands and whitened on the rugged hills of India's north-west frontier; they amazed the Germans at Mons with their rapid rifle-fire and built up a reputation for dogged tenacity amidst adversity in two World Wars. In the beginning the longbow brought the first immortal fame to the common soldier who might otherwise have hardly rated a mention in the history books.

In writing of Crecy, Poitiers and Agincourt one is writing of the Hundred Years' War, of Edward, the Black Prince, and Henry V, but, more than that, one must write the history of the English archer, because without him, and the tactics built around him, none of the victories in France during that mediaeval period would have been possible.

Whether the longbow really altered the course of history is debatable, nor can it be claimed that the English archer contributed towards the foundation of the British Empire. But it cannot be denied that his skill and courage may well have discouraged other, more powerful, nations from attempting to add England to their empires.

Morlaix is mentioned in the text as being one of the earliest actions where the English archer and his longbow demonstrated his almost invincible technique. Here it is, in all its glory, ready to be re-fought as a tabletop wargame.

The Battle of Morlaix 1342 – the first pitched land battle of the Hundred Years' War. In September 1342 an English army of 3,000 under the Earl of Northampton, besieging Morlaix, was suddenly threatened by Count Charles de Blois and a relieving army of 15,000 to 20,000 foot and horse. Abandoning the siege, Northampton marched along the road to Lanmeur to find a position to make a stand. He found it 4 miles (6km) from Morlaix – a small ridge across the road with a forward slope some 300yd (270m) long. A wood to his rear, ideal for hiding his baggage wagons, also prevented him from being out-flanked by cavalry. The English formed up in a line 600yd (540m) long and 50yd (45m) from the wood, with dismounted men-at-arms in the centre and archers on the flanks, and dug a trench as an obstacle to cavalry in front of their line.

In three successive columns, the leading one formed of dismounted local levies, the French advanced straight at the

Where the English stood at Morlaix, taken from just in front of their line where the trench was dug. The wood in the rear is that into which Northampton retreated

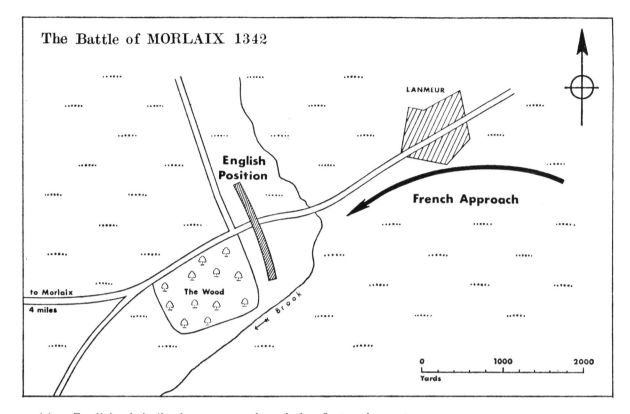

The Battle of MORLAIX 1342

LANMEUR

English Position

French Approach

to Morlaix

4 miles

The Wood

Brook

0   1000   2000

Yards

waiting English. A hail of arrows reduced the first column to disorder, and the second, of mounted men, was stopped by the hidden trench, where a confused mass of men and horses piled up, an easy target for the English archers. A few horsemen succeeded in remounting and crossing the trench, but they were all brought down by arrows.

Now followed a quiet pause while the surviving French retired and the third column prepared to advance. It was larger than the whole English army, extending beyond the English wings and threatening their position from both sides. The two leading columns, each greater than his own small force, had been repulsed, but Northampton knew his archers to be short of arrows, though they had recovered all they could; and the trench was now filled with bodies and presented little obstacle to determined men.

As the third column began a ponderous advance on foot, Northampton withdrew his men to the shelter of the woods, forming a defensive line along the edge of the trees. The archers reserved their scanty ammunition until the French came close, and then fired sparingly but effectively. Not a single Frenchman penetrated the position at any point, although they swung round the flanks and almost surrounded the wood.

In fact Northampton's worries were minimal compared to those of Charles de Blois, whose men, including a number of Genoese mercenary crossbowmen, were deserting on all sides. With night approaching, he abandoned the contest and withdrew

167

A wargames of the Battle of Poitiers 1356. 30mm 'flat' figures (Bath Collection) American Revolution 1776

Formal warfare

Formality was one of the main characteristics of 18th century warfare. The 'Order of Battle' was a formal, laid-down placing of regiments which controlled their positioning when camping, marching and deploying on the battlefield. This positioning of units was mandatory and the displacing of a regiment was a humiliation that could hardly be borne; commanders so displaced were known to ignore their new position and form-up in what they considered to be their honourable place in the ranks of the army. If realism and accuracy is to be maintained when wargaming in the 18th century then an awareness and observance of these rules is essential.

On the field of battle the army was split into the right and left wing, each consisting of one or more divisions (or two or more brigades), which were composed of two or more regiments or battalions. The posts of honour were:-

1. The right of the front line.
2. The left of the front line.
3. The right of the second line.
4. The left of the second line.

Sometimes the advance guard would be in the number 1 position or, when retreating, the rear guard would be so positioned. The first brigade would be on the right flank of the right wing; the second on the left flank of that wing; the third on the right of centre; the fourth on the left of centre and so on. On the left wing, the same order would be reversed, with the first brigade on the left flank, etc. The *continued*

slowly to Lanmeur, at which Northampton formed his small band into a defensive formation and left the wood to return to the siege of Morlaix.

Reconstructing the Battle

This was strictly an affair of firepower, with not a single instance of a blow being struck in hand-to-hand combat. As long as the rules give due weight to the effectiveness of the English archers, the events of 1342 cannot but be repeated on the tabletop battle-field. The lowering of morale caused by losses, together with the already poor morale of the French local levies, will ensure this. The French contributed to their own downfall by attacking in three separate divisions, so that never more than 5,000 or 6,000 of them at any time threatened an English force, 3,000 strong, with extremely high firepower and operating for some of the action from concealment.

As the French fought in successive 'battles' it is not necessary to provide figures for all of them (though the purist wargamer may cavil at using local levies to represent men-at-arms, and vice versa).

Commanders' Classification

Northampton, on the day, was well Above Average. Charles de Blois was well Below Average.

Number and Quality of Men and Style of Fighting

The total force besieging Morlaix is said to have been 3,000, comprising 2,000 men-at-arms and 1,000 archers, which means that the French divisions were turned back by only 1,000 men. De Blois led an army of 15,000 to 20,000 foot and horse, split into three large columns, each of 5,000 to 6,000 men. The leading column was formed of dismounted local levies, ill trained and of low morale; the second was formed of mounted men; and the third of armoured and dismounted men-at-arms.

The French had with them a body of Genoese mercenary crossbowmen, who were expensive professionals, but it is unlikely that they amounted to more than 500. They did not play a very prominent part in the battle, perhaps because, as at Crecy, they were not given very much chance to do so, or because, as they advanced, they were overwhelmed by the weight of archery fire turned upon them.

The result of the Battle of Morlaix failed to convince the French that it was almost hopeless for cavalry to make a frontal attack on a position defended by men-at-arms supported by archers. Alone the superbly trained English archers turned back the three successive French columns, but had one or other of these columns come to grips with the English (as at Agincourt), it is quite likely that they would have been driven back by the combined efforts of the dismounted men-at-arms (well trained professional soldiers) and the nimble archers, now laying about them with sword and maule. The French local levies were frightened and unwilling peasants, probably only armed with sticks, clubs or scythes. The mounted French in the second column would have been heavily armoured knights or men-at-arms, professional soldiers with training, discipline and skill, though their headlong charge at the enemy was now outdated. The third column was composed of similar professional soldiers, though this time they fought on foot.

The French military system in the mid-14th century was the epitome of feudalism, with its lords and their retainers jealous of each other and unwilling to co-operate. Tactical control, therefore, was almost impossible. The mounted knight and man-at-arms considered themselves the masters of the field, and the infantryman was scorned and inadequately employed. Forced to attack by the English method of waiting in carefully chosen defensive positions, the French knights were immobilised when their horses were killed by the archers; and when they attacked on foot, their infantry, equally vulnerable to arrows, reached the English position too exhausted to be effective.

Morale

Made aware of their potential at Halidon Hill some nine years earlier, the English archer/men-at-arms combination, led by exceptional commanders, produced the highest possible state of morale. The French, superior in numbers and seemingly the pursuers, were probably of equally high morale at the onset of the battle, except for their peasant levies whose morale must inevitably have been low.

Terrain

As in almost every battle of the Hundred Years' War, the English

same pattern was followed within brigades for arrangement of regiments, who were given numbers to determine their priority. After the Order of Battle was completed, the artillery deployed along the front line, while cavalry and light troops held the flanks or were deployed as skirmishers out front.

When two or more countries allied together, they had to arrange between themselves the positions of honour. As may well be imagined, this frequently caused such a deep sense of grievance and injustice as to jeopardise the fighting qualities of the allies.

A commander had very little freedom in the actual deployment of his troops on the field of battle – which should be reproduced in wargames of this period. To give the commander a certain degree of flexibility, Grenadier and Light Infantry companies were detached from their regiments and formed as special battalions under his personal command; as these flank companies held the first and second places of honour in their regiments, the battalions formed from them could claim precedence over any regiment and take whatever position the commander selected for them.

169

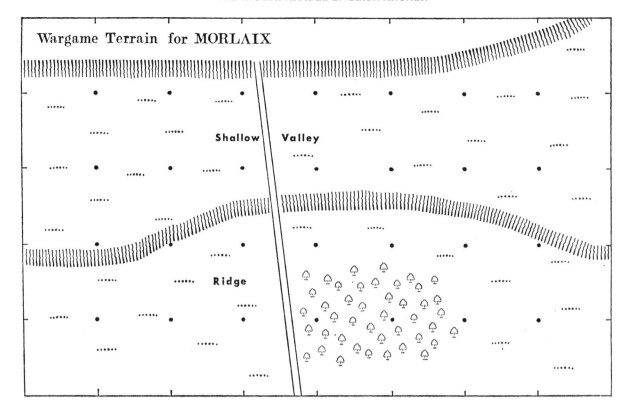

Wargame Terrain for MORLAIX

Shallow Valley

Ridge

position in Morlaix was all important. It was a copybook defensive position, suited to the weapons of the day, and made even more impregnable by Northampton's retreat into the wood, whence his men could fire from concealment.

Military Possibilities

1. The French catch the English before they have taken up a defensive position. This is unlikely, because the latter were able to dig a trench, indicating that Northampton had reached his selected position with time to spare.

2. The French alter the order of their attack, first sending in the dismounted men-at-arms backed by the cavalry, and keeping the unreliable local levies in reserve for mopping-up purposes.

3. The French attack en masse, sending the cavalry round one side of the English position and the local levies round the other, so as to stretch the small English force. Even when in the shelter of the woods, the English were still vulnerable to infantry infiltration through flank and rear.

As in many battles of the Hundred Years' War, the English archers fired so rapidly that they ran short of arrows. Had they run out completely, the French might have come to grips and their overwhelming numbers prevailed. But the Count de Blois's lack of control indicates that he would not have been able to take advantage of even such a favourable situation.

4. Northampton does not retreat into the wood. There is a possibility that his men might have been overwhelmed in the open by the large third column.

CIVILIANS IN WARGAMING

Once upon a time conditions controlling a wargame forced half an army to be employed rescuing the Captain-General's mistress from behind enemy lines – it was a long time before the resulting dislike of civilians in wargames died away. Commonsense indicates there *must* be a place on tabletop terrains for these itinerant human beings who all look different, move stubbornly at their own slow pace, and refuse to acknowledge the benefits of military discipline.

Enough has been said to indicate the importance of the role of the civilian in warfare at all periods of history – yet the wargamer despite seeking realism, blandly and blindly goes his merry way without bothering about the civilians. Consider those farms and villages painstakingly constructed on wargames tables, fiercely fought-over without the inconvenient presence of their inhabitants? Except in the best-organised campaigns, how are our lines of communication through enemy territory so peacefully maintained without withdrawing a single regiment from the much more interesting sharp end? It is almost impossible to find any corner of the world, however seemingly desolate, over which to fight a war where there are NO inhabitants whatsoever – any old 8th Army man will wonderingly tell of the sudden appearance of Arabs in a hitherto barren waste, offering to barter eggs for cigarettes!

For the sake of realism, the civilian population inhabiting the area of a tabletop battle or campaign MUST be allowed to have a certain effect upon operations. This should not be regarded as a bore or a nuisance, despite the consternation caused when tactical skill has brought your battalions onto an enemy flank, and when your firing-turn arrives, you are thwarted by the farmer and his family

A quite superb Peninsular War period. Spanish town made by Peter Gilder, with its civilian inhabitants. 25mm (Figures – Connoisseur, Courtesy Duncan Macfarlane, Wargames Illustrated*)*

running between you and the target! If only because it wastes ammunition, few commanders will order their men to fire under such circumstances. Or how about that herd of cattle stampeding and disorganising your lines – even warrior-Zulus at Rorkes Drift had their edge blunted thus! And what of the civilian/soldier in those conflicts where there is a nebulous dividing-line between a terrorist and a patriot, where men run through the front door as soldiers and emerge at the back as civilians? If they are Orientals, to Western eyes they all look alike anyway! In Vietnam, there was an engaging enemy habit of chubby children sidling up to a GI, to press into his wondering grasp a hand-grenade with the pin out! All this creates considerable alarm and despondency in real-life, so it should certainly open up new vistas for the blasé wargamer.

The civilian role in warfare can be classified thus:

1. The Helpless – refugees, mostly old people, women and children, who are a drag on resources, get in the way when fleeing in panic and generally impede military mobility. In wargaming they represent a frustration-factor.

2. Younger people of both sexes, for political reasons basically belligerent; often resorting to force through sheer desperation. From their ranks come resistance groups, partisans, guerillas – and terrorists. Films like *The Magnificent Seven* or *The Seven Samurai* show how local defence-groups are formed. It does not take too much imagination to realise the scope of this in wargaming.

3. Those subjugated or allied civilian populations who are enlisted into units of militia, to police back-areas or even play an operational role to release recognised military formations for other duties. Such groups vary from the pathetic to the highly efficient, depending upon if they are recruited from recognised warrior-races, such as Gurkhas, Sikhs, Zulus, N.W. Frontier tribesmen, or the ferocious Swiss peasants and their pike-formations. Many spring to mind – Lonsdale's and Bettington's Horse in the Zulu War of 1879; Sikh police and mercenary levies in India; and those white expatriates who made-up exotically-named units in various Colonial areas. Bear in mind that history has many examples of all three classes eschewing violence to become a universal obstacle through a policy of passive resistance, masterminded by a Gandhi-like figure.

Unless driven-out with retreating enemy forces, a civilian population will inhabit the occupied areas – they may be hostile, part-hostile, apathetic or downright submissive – most likely, a tantalising blend. In any event, they have to be supplied and controlled by the troops of a military government, who guard points of strategic importance and maintain lines of communication, putting an obvious strain on both man-power and supply-services. It is suggested that when setting-up a campaign, towns and villages should be listed and each given a scaled-down population in the same way as a regiment equates to real-life formations by having one figure representing say twenty men. Rule that one soldier has to be detached for every twenty civilians, to police them and perform necessary occupational duties. This proportion can vary in the event of resistance when guerilla and partisan activities have to be combated; troops will have to be withdrawn in greater numbers from forward forces, thus weakening them. As it is known that hostile activities are encouraged by weak or non-existent occupation forces, rule that when occupational forces fall below one soldier for every twenty civilians, your opponent may allocate an agreed number of his civilians to form partisan bands, operating on your lines of communication as loosely-organised groups with twice the mobility of Regular units, but with only half their fire and melee-power.

Refugees should be given a relatively important status; they block roads and hampering troop-movements; this can be turned to advantage by having them driven down roads along which the enemy are advancing, thus slowing him down. Troops must be detached to cope with this, say one soldier being able to drive before him ten civilians at half-normal rate of movement.

Campaigns can include Standing Orders – "ALL civilians and their livestock in outlying farms, towns and villages will be evacuated under adequate escort to places of safety. Should this not be possible, these places, including single farms, will be defended by bodies of troops detailed for such purposes. Under no circumstances will civilians or livestock be allowed to fall into enemy hands". Rules insist on civilians and their cattle moving at a very slow rate, limiting escorting military forces to the same speed; and each group of civilians have to be protected by half their number of soldiers. A situation is recalled when a Commander displayed a callous disregard for the civilian's welfare, hastening his troops to the zone of operations, leaving the unfortunate settlers in their outlying farms to fend for themselves. Subsequently it was ruled that for every individual civilian, cow or sheep killed or captured through such neglect, one man was deducted from the actual numerical strength of the offending commander's force.

Turkish women fighting alongside their men against the Russians in the war of 1877 – probably at the siege of Plevna

What made fearful peasants, men not of arms but of the soil, fight back against Viking invaders? Desperation was the name of the game – the emotion causing caution to be cast to the winds, turning peaceful men into tigers stimulated by awareness of the heavy price of defeat, they resist far beyond normal or expected capabilities, fearfully aware of the only alternative to be death or torture for themselves and their families. At the siege of Rouen in 1418 when the last dog and cat had been eaten and fifty thousand had died of starvation the long-suffering inhabitants rose in revolt against the magistrates, demanding that negotiations be opened with Henry V. Ignored, the civilians desperately fired their own town in several places, undermined a length of wall and supported it with props, suddenly removed when the whole population were to sally forth and endeavour to fight their way out, women and children making their escape in the confusion. It so happened that their desperation was not needed as Henry relented, fearing the destruction of Rouen, the largest city of Normany.

Simulated in its simplest form, dice-throws can be made to decide if the worm turns and if a desperation-factor is present; give values to those things that make a man desperate, adding to the relevant total a random-dice score, the result being decided according to the gradings on a chart. Perhaps Desperation-Factors should only be evoked against enemy hallmarked by known Ferocity – such as Assyrians, Mongols, Huns, Vikings, Swiss, Landsknechts and such barbarians as Red Indians, Zulus and North West Frontier tribesmen; in our own times SS units, the Japanese and some of the the elite specialist formations.

On D-Day French Commandos attacking the Casino at Oustreham were greatly aided by local Frenchman M. Lefevre who, after cutting power-lines to German electrically-fired flame-throwers protecting the Casino, accompanied the Commandos advising them where enemy positions were strongest. Such a brave and helpful civilian might be introduced into other operations, or instead of being an ally he could be a quisling leading the raiders into a trap.

Such glib encouragement of the use of civilians in wargaming must be tempered by the availability of actual models – in past halcyon days cheap boxes of plastic OO/HO scale figures were both a boon and a blessing – what unlimited conversions could be perpetrated upon the inhabitants of such boxes as Modern Civilians; Robin Hood; Wagon-Train; Wild West cowboys and Indians; Tarzan; Gold-Miners; Prospectors; Wild West camps and towns, Sheriff and Outlaws, etc. The farm animals, mules and bison provided the necessary livestock.

Today, these plastic figures are hard to find although they still exist; but other manufacturers turn out civilians and probably conversions can be made from the incredible ranges of fantasy figures listed in our journals. The widest variety of civilian figures can be found in the catalogues of model railway accessories, particularly those of Continental makers like Preiser of Germany whose elaborate 'Katalog' lists hosts of civilians in many scales. There are every conceivable type of artisan, workman, animal, man, woman, children – even wedding parties, firemen, circus-performers, tourists of many nations etc, etc. They come in a variety of scales – 1:87; HO; 1:72; 1:32; and 1:22.5. They can be had painted or unpainted and, of particular interest to wargamers, there is a range of civilians in dress of 1835 and 1860.

Undoubtedly, the course of many historical campaigns were altered or determined by the actions and reactions of civilians. If true to avowed principles of seeking realism, the wargamer must include these irritating, undisciplined, untidy and unmanageable members of the human-race in his tabletop operations. Let there be no more deserted landscapes, no more villages bereft of inhabitants – even in the best-ordered circles there was always Gaffer Giles or Aunt Mabel who were too infirm or too obstinate to move out in the face of occupying troops – make room for them on your table!

IT CHANGED THE COURSE OF HISTORY!
The War of the American Revolution 1775–1783

An officer of Baylor's Dragoons (American)

There is a stirring ring about the names of battles of the American Revolution – King's Mountain, Freeman's Farm, Bennington, Brandywine, Monmouth, Cowpens, Guildford Courthouse – that makes re-fighting them a stimulating proposition. And it is one that can be practically achieved with a considerable degree of realism because most of them are highly suitable for transferring to the confines of our wargames tables – mostly because they involved small numbers so that realistic scaling-down is possible. For example, Brandywine in 1777 was one of the biggest battles of the entire war yet Washington's army totalled only 10,500 men, and at Camden in 1780, Gates with 3,000 men was driven from the field by Cornwallis at the head of a mere 2,400 – yet by the standards of this war it was a major battle. Most of these late 18th century battles in North America were contests of manoeuvre rather than of fighting ability, largely because of the lack of artillery plus the rapid and accurate musketry which indicated the vital role to be played in European warfare by the rifleman and his rifle. Frontiersmen, such as those led by Morgan, operated independently in groups. Their long rifle, proven in many Indian wars, was a vastly superior weapon to the smoothbore musket which could only fire with 40 percent accuracy at 100yd (90m) against the rifle's 50 percent accuracy at 300yd (270m). By using a greased patch around the ball range, accuracy and penetration were improved. However, it took twice as long to load and, unable to carry a bayonet, was useless against a bayonet attack and could not stand ground when rapid fire was required. When Burgoyne attacked the Americans at Freeman's Farm it was a tactical landmark because Morgan's Riflemen bore the brunt of the battle – in brown linen hunting shirts and buckskin breeches, almost invisible amid the Autumn colouring of the woods, they took heavy toll of the scarlet uniformed British infantry, advancing in close-order across forest clearings.

When fighting took place in the wilderness where the frontiersmen were notably better in their use of cover, American irregular tactics were superior to those of the British and Hessians. Under the same conditions, the British Light Infantryman was probably superior to the average American soldier. But the Revolution could not be won in the Wilderness, so American soldiers adapted to current military thought and tactics.

With the possible exception of the Prussians, the British

Line Infantry in the American Revolution could shoot faster and more accurately than anyone else. One American problem was to train their infantry to stand and trade volley for volley with the British, particularly when fighting in the settled regions along the coast where European tactics prevailed. Washington was defeated at Brandywine and Germantown, where it was shown that short-term American volunteers were not yet ready to meet the Regulars in formal combat, falling easy victims to bayonets that were on them before they could reload. The musket had a 20 percent hit probability at 150yd (135m), but infantry could charge 180yd (160m) in a minute so that troops close enough to suffer casualties from musketry were also within charge range. The psychological effect of the bayonet brought the British many victories during the American Revolution.

The brigades of the Continental army were usually formed of regiments of one State or closely related States. They varied in strength, some being equal to two regiments and others only having the strength of a single company, while many commanders organised their troops to suit themselves so that some companies were larger than other regiments. Often, they were so much below strength that one or more line companies would be dissolved and absorbed into remaining companies. Organised to resemble the British system, the paper strength of a Continental regiment was nine companies, each of eight officers and sixty-eight men, eight being Regular infantry with the ninth an elite company of Light infantry. To create a standard-size unit, prevent an irregularly formed line and make brigading easier, Von Steuben decreed that a battalion should consist of 200 men so that one regiment could form two or more battalions while a number of the company-size regiments could form another. When possible, militia were organised along the same lines although it was difficult because of the unpredictable number of militia who obeyed the call. The usual practice was to group militia in one large formation in the immediate vicinity of other militiamen from the same State.

Artillery was still a relatively minor arm during the American Revolution with hired civilians inefficiently and unreliably transporting the guns in and out of battle. The largest American artillery to operate as a whole was a company of four guns and sixty artillerymen. Never very plentiful, American artillery pieces were individually placed to cover important fields of fire. With experience, American artillery became proficient and had probably attained parity with the British by late 1777.

British infantry battalions of the American Revolutionary period consisted of ten companies – one grenadier company, one light infantry company and eight centre companies usually totalling about 400, although the paper strength was 475 men. A Hessian fusilier or musketeer battalion had five companies – one of grenadier and four of fusiliers or musketeers. The paper strength was 650 but casualties and desertions considerably depleted most Hessian battalions. British Line infantry battalions

American Revolution 1776 – A Pennsylvania Rifleman

American Revolution 1776 – A Hessian Musketeer

175

American Revolution 1776 – Sherburne's Continental Regiment skirmish with British Light Company men

had attached to them a light field piece manned by an NCO and a dozen privates; sometimes these guns were grouped together to form a massed battery. It was the practice in both British and American armies to detach light companies from their battalions and form them into 'elite' brigades; the British also 'un-brigaded' the grenadiers, light dragoons, the guards and sometimes the Highlanders to form detached forces, like the flank companies at the personal disposal of the commander.

Accounts of the battles of the American Revolution reveal the varying, usually poor, quality of the American militia units. When wargaming, this can be simulated by allocating to such units a 'panic' rating of 5 points; to cause panic in the units this figure has to be doubled by adding a dice score to the 'panic-triggering' score of 7 points allocated to a Regular infantry unit or the 8 points of a cavalry unit. A direct bayonet charge by infantry or sabre charge by cavalry will cause 1 point to be deducted from the militia score. Thus, if a Regular infantry unit throws 2 which added to their 7 points makes a total of 9, then it is insufficient to trigger off panic; but if they throw 3, this added to their 'panic-triggering' figure of 7 points will cause the militia to turn and run, at charge-move speed, in a direct straight line to the nearest enemy-free edge of the board.

The availability of inexpensive HO scale plastic figures, made by Airfix and ESCI, brings this fascinating period within reach of even the most poverty stricken wargamer. It has many unique aspects that make it well worth fighting – and most of them are revealed when fighting.

176

The Battle of Guilford Courthouse 15 March 1781

This encounter took place during the American Revolution (1775–83) between an American force under General Nathaniel Greene and a British army under General Cornwallis. The contrast between the rigidity of the British Regular forces and the loose and unorthodox tactics of the American Militia makes for unusual but fascinating table-top battling.

This British victory marked the beginning of the end of the Revolution, because, by winning, Cornwallis so weakened his army that he lost the campaign in the southern colonies. Cornwallis had been pursuing Greene in the Carolinas in the hope of forcing an engagement, while Greene, wishing to avoid battle, was trying to draw the British as far as possible from their base. Early in March Greene collected reinforcements in Virginia that strengthened his force to about 5,500 men, moved westwards to Guilford Courthouse, within 12 miles of the British Army, and took up battle stations facing west in three lines 400yd (366m) apart. At the head of a force of 1,900 men with three guns, Cornwallis marched at daybreak on 15 March. About 4 miles (6.4km) from Guilford Courthouse the advance cavalry of both armies met in a brief skirmish, and the Americans were driven back. Continuing the advance along the New Garden Road, the British crossed a small stream from which the ground rose gradually to an open space about 500yd (460m) square, made up of three cultivated fields bordered by rail fences; the whole

American Continental line infantryman

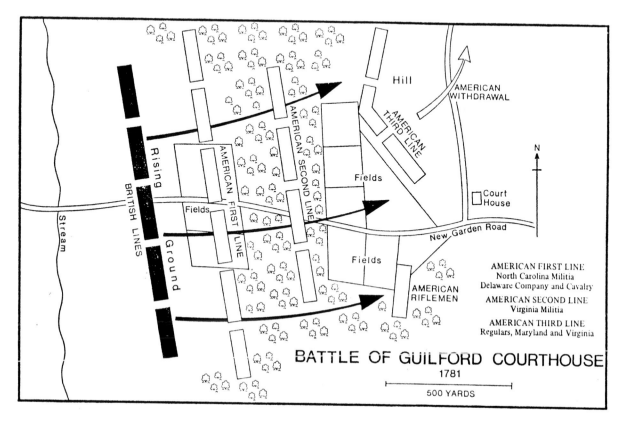

BATTLE OF GUILFORD COURTHOUSE
1781
500 YARDS

AMERICAN FIRST LINE
North Carolina Militia
Delaware Company and Cavalry

AMERICAN SECOND LINE
Virginia Militia

AMERICAN THIRD LINE
Regulars, Maryland and Virginia

American Revolution 1776 – An officer of the British Grenadiers

A Sergeant of the Green Mountain Rangers

area forming a defile between thick copses of trees. Here, with an excellent field of fire and protected by the rail fences, Greene had drawn up his first line of 1,600 men. This line consisted of two brigades of North Carolina Militia, untrained soldiers without battle experience; two battle-experienced units – Lee's Legion (Regulars) and Campbell's Riflemen (frontiersmen from Virginia and the North Carolina mountains) – on their left; and Colonel William Washington's Regular Cavalry, some of the Delaware Regiment of Continentals and Lynch's Riflemen (veteran frontiersmen) on their right. Two 6pdr guns in the centre on the road commanded the stream crossing.

About ½ mile (80m) ahead of the clearing, in the woods that closed in on the road, Greene deployed his second line of about 1,000 Virginia Militia, untrained and inexperienced. Commanded by men who had served in the Continental army and had some battle experience, this second line was somewhat stronger than the first. All the troops of the second line were hidden in the woods on each side of the road, with picked marksmen behind them, ready to shoot any man who ran away.

Then came another open space of cultivated ground, made uneven by hollows, and 400yd (366m) in rear of the second line, and rather to the west of the road, was a hill that formed the salient angle in the midst of the clearing around the Courthouse. Near the eastern edge of this clearing Greene drew up his third line of 1,450 Regular troops, the 2nd Virginia Brigade on the right and the 2nd Maryland Brigade on the left, with two 6pdr guns between them. The right flank was unprotected, but the left flank, resting on the New Garden Road, was protected by artillery during the later stages of the battle. Both flanks of the first two lines were unprotected but as the heavy woods forced the British to attack frontally, these exposed flanks were not a disadvantage.

Cornwallis' army was much better organised, disciplined and trained than the American, and his men, perhaps the best British forces in America, were veterans commanded by able experienced officers. Their right wing under Major-General Leslie consisted of Bose's Hessian Regiment and Fraser's Highlanders in the first line, with the 1st Guards in support; the left wing was formed of the 23rd and 33rd Regiments under Colonel Webster in the first line, with the Grenadiers and the 2nd Guards in support. A small corps of German Jagers and the Light Infantry were stationed in the wood to the left of the road, while Tarleton with the cavalry remained in the rear on the road. Cornwallis posted his three guns on the road itself; they could not move anywhere else because the woods restricted their field of fire.

The British force formed up under American artillery fire after crossing the stream at the foot of the hill, losing very few men, while the British artillery replied with an equally useless expenditure of ammunition. At about 1.30pm the British troops advanced across the first clearing, amid a hail of fire from the invisible enemy in front and on the flanks, then charged with the bayonet. The North Carolina Militia melted away in panic, while Greene's two guns retired to take post near the road with the American third line. The British line halted and the flank battalions turned outwards to oppose the American riflemen, who were still pouring in a destructive fire from both sides. Bose's regiment wheeled off to the right flank, and the 1st Guards moved into the line to take their place next to the Highlanders; on the other wing the 33rd, with the Jagers and Light Infantry on their left, similarly wheeled off to the left flank and the Grenadiers and the 2nd Guards advanced to the left of the 23rd.

Slowly the redcoats pressed forward through the trees, driving the American riflemen back at the bayonet point; the cavalry of Lee and Washington fell back to conform. Bose's Regiment and the 1st Guards drove Lee and Campbell's left flank detachment backwards in a south-easterly direction in a struggle that completely detached them from the main course of the battle, their separate engagement only being broken off by the Americans at about the time the main conflict ended. The American right flank detachment briefly took position on the flank of the second line, and, when that retired, moved to the flank of the third line.

The attack proceeded east along the road and through the wood for about 400yd (366m) until it struck the second line, where the rifle-armed Virginia Militia stubbornly held their ground, inflicting heavy losses on the advancing redcoats. Discipline and experience began to tell, and gradually Wilson's Brigade on the American right crumbled before the Light Infantry and Jagers, who were their equals at bush fighting. Stevens' Brigade, on Wilson's left, stood firm for a while, but finally the pressure told and both forces were pushed back until, reaching the road, they broke and retired in disorder through the wood to Greene's third line, protected from pursuit by Washington's cavalry retiring with them to the edge of the forest.

Entirely north of the road, Greene's last line was opposed by the British left wing, which was slowed up by heavy undergrowth and deep gullies. Led by Webster, the now considerably reduced 33rd, Light Infantry and Jagers, attacked the 1st Maryland Regiment, the finest battalion in the American Army, which steadily awaited the assault before pouring in a volley at close range and charging with the bayonet to drive the attackers back in disorder, with heavy loss. Although severely wounded, Webster drew his men off into the shelter of the woods and rallied them, supported by the three British 3pdr guns, which advanced along the road and unlimbered on the rising ground at the edge of the forest, whence they kept up a steady and well directed fire.

While Webster's force was rallying, General O'Hara led the 2nd Guards and the Grenadiers, supported by the 23rd and the Highlanders, against the Maryland Brigade. Seemingly giving way before O'Hara's attack, the 2nd Maryland Regiment steadily withdrew as the Guards pressed them; suddenly Washington's cavalry galloped out of the woods to crash into the rear of the Guards, while the 1st Maryland Regiment flooded out of the

American Revolution 1776 – A drummer of an American Continental line regiment

American Revolution 1776 – A Hessian Grenadier

179

The Spanish–American War 1898

By modern standards this was a small war and cannot be considered as a typically Colonial conflict because white men with modern weapons fought on both sides.

The Americans used batteries of Gatling-guns and, without much difficulty, succeeded in winning the war. It was a Naval conflict as well as on land – the American fleets quickly isolated Cuba, Puerto Rico and the Philippines and later completely destroyed the Spanish fleet without loss to themselves.

From the point of view of a semi-modern conflict, on a small scale, fought amid jungle-like terrain, the Spanish-American war can provide a fruitful source of pleasure to the Wargamer.

GUILFORD COURTHOUSE

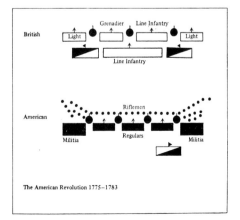

The American Revolution 1775–1783

undergrowth on to their left flank. Fighting fiercely, the Guards were utterly broken, and the danger of a wholesale retreat forced Cornwallis to order his artillery to fire grapeshot into the struggling mass, causing great losses to both Americans and British.

In the pause that followed Cornwallis reformed his line, rallying the 2nd Guards, who had been joined by the 1st Battalion; Webster led up the 33rd, the Jagers and the Light Infantry on their left. Under the smoke of a volley Tarleton's Cavalry were sent in, to end the American resistance on the British right and then they moved across to join Bose's Regiment and press upon the Americans' left flank, while the remainder of the British line engaged their front.

Seeing that the day was lost, Greene ordered retreat, abandoning his guns; there was no pursuit by Cornwallis' exhausted troops. Casualties were 93 British killed and 439 wounded; 78 Americans killed and 183 wounded.

Rating of Commanders and Observations

Greene undoubtedly displayed stratetical sense by fighting not to win but to cause Cornwallis such losses as to diminish his future prospects of victory. Greene is certainly an 'average' commander but could be classified as 'above average' here, though that rating would probably put victory completely beyond the reach of Cornwallis (himself an 'average' commander), with his

180

considerably smaller army. Webster could be 'above average', as could Washington; and other commanders can be classified in relation to the performance of their units. It will be necessary to assess the varying states of morale and fighting ability of

(a) the British Regulars,
(b) the Continental Regulars,
(c) the American Riflemen, and
(d) the American militia.

This assessment could be handled after the manner of the Wargames Research Group's Ancient Rules, where Regulars, Barbarians and levies, etc, are given different values when

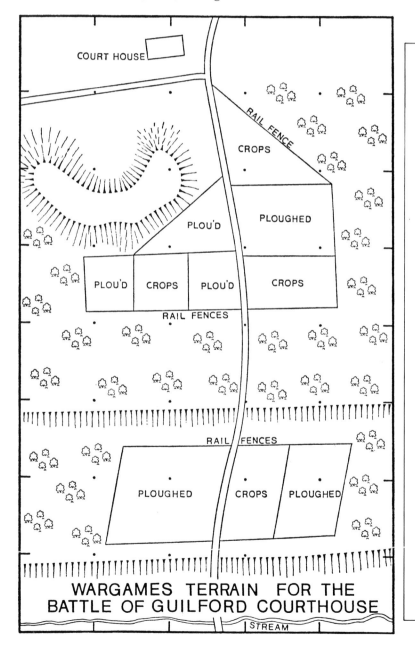

WARGAMES TERRAIN FOR THE
BATTLE OF GUILFORD COURTHOUSE

Plan a Wargame No. 3
Fighting a wargame with more than one player on each side has certain built in snags. In a completely unrealistic manner, the two allied generals confer together in a conversational fashion as they make their decisions; amid the real fog-of-war neither would have much idea of the other's activities and frantic messages would be sent back and forth. Also, these games invariably seem to develop into two completely separate battles in which each quarter of the table fights its similar opposing sector.

A better and more realistic method is to fight the battle in two sections, by having one general on each side in command of the Advance Guard, whilst the second general (commanding the Main Body) remains away from the wargames room and sees the terrain for the first time only when his troops arrive on it.

The Generals commanding the Main Bodies will have *no* information other than that contained in despatch and map. They will not see the terrain beforehand and they will list (in writing) their dispositions on coming onto the field *before* they see the table.

building up an army. Otherwise, the British Regulars and the Continental Regulars could be considered as top-grade soldiers; the Riflemen classed as light infantry, but not given the high morale of the Regulars, because of lacking the Regulars' cohesion; and the militia given a low grading, both in morale and fighting ability, because of the likelihood of their fleeing the field without firing a shot.

With the numerically superior American army in positions of their own choosing and the British historically pledged to attack in rigid formations, the only manner in which this battle can be accurately simulated is to devise 'local' rules that balance these factors. The Militia must check their morale state as soon as the British come within charge-move distance, whenever they take fire and sustain casualties, or whenever they are brought into melee contact. The British Light Infantry and the Jagers should be given the same powers of manoeuvrability as the American Riflemen.

The cohesion of the British units attacking in rigid formations should give them an increased morale bonus, or else it is doubtful whether they will ever get into melee contact, as they did in real life. In fact, the morale of the British should be capable of being raised to great heights; particularly in the case of the Guards, who were driven off after being severely mauled by the high-grade Maryland Regulars, hit in the rear by cavalry and then flailed with grapeshot by their own artillery. They subsequently rallied and came forward again to the attack – a rare exploit, unlikely to occur under normal morale rules.

Military Possibilities can be devised to prevent the Militia from fleeing, to prevent the American second line from being pushed back, to alter the withdrawal of Webster's men and even their subsequent rallying, and to allow the Guards to beat the Maryland Regiments and to turn and fight off the cavalry rear attack, etc, etc. But the historical reality of Greene's delaying action to maul his opponent, who consistently pushed forward, indicates that the most realistic simulation will be achieved by judicious handling of the morale situation.

Construction of Terrain

This battle, being fought in three separate stages, requires the extensive area allowed by laying the terrain lengthways, with the stream on the 'bottom' baseline. From there a ridge, 18in (45cm) wide on its top, stretches right across the table, with three rail-fenced fields, each about 18in (45cm) square. Because fighting has to take place amid the trees, the comparatively spacious areas of woodland must be made 'symbolic', by representing them with hardboard bases bearing occasional trees around their edges, the whole being the wood itself. The extensive ridge and hill are easy to form by draping grass-coloured cloth over shapes below. Considerable lengths of rail fence are required; this can be made from basket-weaving cane, or the thin wire used for tying plants.

THEY FOUGHT FOR GOLD!

Mercenaries of the Middle Ages

From the beginnings of the recorded history of warfare, men have sold their fighting services to the highest bidder. Sometimes they did this as individuals and on other occasions they banded themselves together into Free Companies and Condottierri. From a wargames point of view the employment of such fighting men will probably apply more to the ancient period – the suggestions made below are largely confined to that age.

Mercenary troops were not always entirely trustworthy and the chance of their mutinying must be considered. When a commander is facing mercenaries in battle he has the right to challenge them once. He throws one dice for the entire Free Company – a 1 means that they will not fight for their employer and a second dice throw reveals their course of action. A 1, 2 or 3 means that the mercenaries change sides and fight against their former employer, a 4, 5 or 6 means that they march off the battlefield by the shortest route. If they change sides the new employer must pay them double wages.

When a mercenary force is destroyed in battle it simply goes back to the central pool and is available for hire in the normal way on future occasions. If a mercenary force is captured during battle their captors can throw a dice and a 4, 5 or 6 throw will mean that the mercenaries agree to change sides, a 1, 2 or 3 means that they are returned to the central pool.

In some historical periods, battles were fought with more of an eye to material gain than to the actual destruction of the enemy. If the lower ranks of a defeated side know they were going to get their throats cut then with equal certainty their feudal lords were aware that, on payment of a calculated ramsom, they would be returned to the bosom of their families until the next time they were captured. This business went a step further and, perhaps less magnaminously than it seems, even included the lower ranks when mercenaries and Condottiere took the field. The commander of a Free Company, having taken great pains to recruit skilled professional soldiers, had the most natural aversion towards losing them together with the fee per head he got for offering their services to whatever king or prince paid the most. And when mercenaries met mercenaries, both with the consciousness that today's enemies might be tomorrow's allies and the king you fought against today might be paying your wages next week, then the casualty rate was unnaturally low.

Perhaps the most ferocious and efficient of all mercenaries, the Swiss and German Landesknechts were prominent on many 16th century European battlefields. Here are three of them, ready for wargaming action

To translate this into wargaming terms, a commander in a strong position should be able to offer (within the framework of the rules) his less fortunate opponent the 'Honours of War' where the loser surrenders but retains his honour. One oft-voiced objection to wargaming lies in the fact that there are few if any provisions or indeed inducement to take prisoners – and this might well be the controlling factor in this case. A suggested method is to give each army a points total. In this way the points value of a particular type of soldier when captured would be allotted to his captor and when the 'Honours of War' had been accepted then by negotiation a percentage of that total would be handed back to the surrendering side. For example, say the valuation for a Regular heavy cavalryman is 10 points – say ten such heavy cavalrymen have surrendered, thus giving a total of 100 points to their captors, but as they only surrendered on the condition that half of that total was returned to them then, by accepting the 'Honours of War', the captors have gained 50 points whilst the captives have similarly achieved 50 points. On the other hand, if a belligerent commander refuses the 'Honours of War' and his men are subsequently

The best known of modern mercenaries, the Foreign Legion Mule Company march across the desert. Scruby 30mm figures

killed then the enemy gain not only the points of value of each man killed but also a bonus of 50 per cent of that total. Or it may be considered sufficient for the defeated side to lose the total points value whilst the triumphant army gain it.

From these suggestions arises the idea of a relatively new conception of wargaming – the man-oeuvre game. Both armies are considered to consist entirely of mercenaries whose command-ers are pledged to avoid killing the enemy's valuable soldiers and, in turn, to avoid losing their own highly priced commodity. Both agree to avail themselves of the 'Honours of War' system when the situation arises for its application. Under normal war games rules, each side tries to man-oeuvre the enemy army or its separate units into untenable positions so that 'Honours of War' can be offered. The points of value of each force are totalled at the start of the game and, on its conclusion, victory goes to the side with the highest points total.

German Landescknechts posed against a tapestry background of the Battle of Granson where, in 1476, their Swiss counterparts achieved a great victory. (Figures – Elastolin 60mm)

ALL YOU NEED TO KNOW TO BE A WARGAMER

No man is an island, and he certainly cannot afford to be one when taking up wargaming, nor does he need to be because there can be few hobbies or interests about which information is so freely available and disseminated. No need to painfully experiment or grub around for vital details when almost all that is required can be found in the pages of the regularly published wargaming magazines, including making social contacts to bring an opponent to the other side of the wargames table. The canny wargamer will seek out the principal magazines *Military Modelling; Miniature Wargames; Practical Wargamer* (quarterly) and *Wargames Illustrated*, subsequently taking one (or all) of these monthly journals on a regular basis – they can be found on the shelves of any large newsagents, such as W. H. Smith.

Besides invaluable well illustrated articles written by experts, each contains stimulating coloured pictures of model soldiers and wargames, but perhaps the greatest value are the galaxy of services offered both via advertisements and in the text. Not only will each magazine put the novice-wargamer in touch with a wargamer or club in his area, but it also informs on forthcoming public conventions and meetings; tells of specific interest groups; lists professionals prepared to paint armies for you; details sources of obtaining ready-painted complete armies; and, through pages of enticing adverts, tells of the makers and suppliers of every conceivable period, scale and type of model soldiers the wargamer could ever dream of possessing. It is even possible to save yourself the trouble of laboriously compiling a set of probably inadequate rules to control your tabletop battles, by sending for a set of professionally-designed rules, for every conceivable period and type of army, plentifully publicised in these pages.

All this merely scratches the surface and these journals justify study because the information they contain makes life much easier for the aspiring wargamer through opening up new avenues and vistas which he would not otherwise know.

Wargamers suffer from lack of time and money – both required in some quantity before gloriously-garbed armies of model soldier manoeuvre magnificently on scaled-down battlefields. And tabletop battles need model soldiers – armies of them – preferably inexpensive, readily available and capable of taking the field with a minimum of preparation. Of course, you get what you pay for and artistically-designed miniatures, although relatively inexpensive for what they are, do not really come cheap. One thing is certain, the beginner may do it 'on the cheap' initially, but with increased experience and interest, sooner or later

Spanish-made 30mm wargames figures – by Alymer – troops of the Directorate, France 1800 (De Gre collection)

Wargame using 1:300 scale figures, revealing how effective these small models can be (Richard Ellis collection)

he will branch out into the best metal figures, his appetite having been whetted on cheaper plastic models.

Undoubtedly the most satisfactory 'cheap' figures obtainable are the extensive selection of 00/H0 scale plastic moulded soldiers, in boxes containing nearly 50 infantrymen for under £2, in a very wide range of periods and types. Initially made by Airfix in U.K., the ever-growing variety appearing regularly in the model-shops quite revolutionised the hobby of wargaming and probably did more to get it off the ground than anything else. At their peak they could be obtained in breathtaking variety, from Tarzan and Robin Hood (with the Sheriff of Nottingham's men as an appropriate enemy) to World War II Airborne and Commando troops of Britain, America and Germany. Their new overseas manufacturer periodically puts familiar boxes on the market. Of late they have been reinforced by a wide variety of excellently moulded plastic soldiers in the same scale, made in Italy by ESCI but readily available in British (and no doubt American) hobby-shops. Currently, this range extends to British Colonial soldiers and enemies (Zulus; tribesmen, and Arabs); Romans and Barbarians; Napoleonic infantry, cavalry and artillery of all the countries engaged in those prolonged wars; Crimean British and Russians; American Civil War Federals and Confederates; and modern American and other troops. Another Italian range with some interesting off-beat (particularly Wild West) figures are ATLANTIC, usually found in small back-street shops; they are still readily available in France.

The majority of wargamers prefer their regiments to be formed of figures all in the same position – i.e. marching, firing, advancing, charging, etc., etc., with an officer or two, a standard-bearer and perhaps a drummer or bugler. When buying metal figures this presents no problem as you order exactly what you want; it is even easier when using 00/H0 scale plastic figures (Airfix or ESCI) as each box is made-up of about 48 figures in a variety of positions, thus there might be 6 men kneeling firing; 6 men standing firing; 6 men advancing; 6 men marching; 6 men running; 1 mounted officer; 2 foot officers; 1 standard-bearer with standard; 1 drummer; 1 bugler; then there are what the experienced wargamer usually terms 'useless figures' – 4 men reloading; 2 men carrying a wounded comrade; 1 man carrying a gunpowder barrel, etc., etc. Thus three boxes can provide:–

Regiment No 1 – A mounted officer; a foot-officer; a standard-bearer; a drummer, and 18 men in the standing, firing position.

Regiment No 2 – A mounted officer; a foot-officer; a standard-bearer; a bugler, and 18 other ranks in the 'advancing' position.

Regiment No 3 – A mounted officer; a foot-officer; a drummer, and 18 men marching.

Regiment No 4 – 1 foot-officer; 1 drummer; 1 bugler; and 18 running men.

Regiment No 5 – 2 foot-officers; a standard-bearer; a bugler and 18 men in the kneeling-firing position.

Five reasonable regiments at about £1 each – that can't be bad!

Never has the wargamer been so well off as at the present time, when there are literally hundreds of model-soldier makers in Great Britain and USA, all turning out beautifully designed and cast metal figures in the currently fashionable scales of 15mm and 25mm. Vying with each other in their some-times quite esoteric choice of historical periods, wars, campaigns and types of model soldiers, they have transformed the manufacture of these minia-ture warriors into an art-form and, both indi-vidually and collectively are so good that recom-mendation is invidious and choice has to be left to the taste of the collector/wargamer. Proliferating daily, innumerable manufacturers and suppliers are known to be operative at the time of writing. It has to be mentioned that some model soldier designers, in their understandable desire to be distinctive, tend to 'adapt' figure-scales, to produce models that appear incongruous when

A selection of beautifully painted 25mm figures of French revolutionary army late 18th century. (Made by Battle Honours by whose courtesy photograph is presented)

fighting alongside comrades reputedly of the same physical size, but designed by a different maker. Sometimes it becomes necessary for the wargamer to restrict his armies to figures made by the same maker, or to carefully seek out other ranges of comparable size and scale.

Few established wargamers mix metal and plastic figures; the majority prefer the heavier and more stable metal figures, but there are innumerable contented wargamers fighting their battles 'on the cheap' with armies of plastic figures, to the almost total exclusion of metal models. With considerable justification, they claim that the pronounced definition of plastic models makes them easier to paint, although conversely the softer material of which they are made renders them less likely to retain the painted uniforms so laboriously applied.

A brief but helpful note on the subject of scales – 25mm figures are about 1in in height; 00/H0 scale (plastic figures) is usually slightly smaller than 25mm but not enough to prevent both units of scales blending satisfactorily on the wargames table. Possibly the most fashionable wargaming scale in use today is 15mm which gives a figure of .590in in height; said to be no more difficult to paint than the larger figures, armies of them present a very pleasing appearance and, of course, allow more troops to be present on the restricted confines of a wargames battlefield than if the larger scale is used. As already mentioned, some reputed 25mm figures more nearly approach 30mm ($1\frac{3}{16}$ in) which allows them to be miniature sculpted works of art – but they are Goliaths among ranks of 25mm soldiers!

Among the most praiseworthy of all model-soldier production lines are the 1:300-scale metal figures marketed by Heroics and Ros Figures. 1:300 scale is equivalent to 1mm = 1 foot (or 3.3mm = 1 metre); a foot figure stands $\frac{1}{4}$in (6mm) tall and a cavalry figure correspondingly taller. At first glance it might seem unattractive to wargame with figures so small that a squadron of cavalry can literally be balanced on a thumbnail! But closer scrutiny reveals that, despite their minuteness, these figures possess a degree of detail equal to that of models five times their size, each infantryman, horse, rider or gun is quite unmistakably what it purports to be. Because their effect is essentially 'in mass', the standard of painting need not be so high and subsequently is much quicker and – devotees claim – far easier than colouring 15 or 25mm figures! Heroics and Ros Figures are readily available in exceptional variety and in packs of 50 infantry, 20 cavalry or 6 guns at an exceptionally low price. It is not possible to list their available range, but any wargamer would be hard put to choose a period of warfare, from Ancient Rome to NATO, that this maker cannot supply. They also make nearly 500 highly detailed model tanks and vehicles in the same scale for World War II period up to the armies of today. There can be no doubt that by using this scale of figure, both authentically and visually, tabletop wargames will more closely resemble real-life warfare than by any other method – that dream of re-fighting Waterloo or Borodino with the actual numbers of historical regiments can come true!

Of course, it is these vast sprawling Waterloo-type wargames with rank upon rank of infantry, charging cavalry squadrons and massed batteries of guns that attract the beginner – then he works out the cost of it all and seeks other methods of fulfilling such dreams! It has already been told how it can be done with the very smallest model soldiers, and another way is to make one's own figures, casting them in metal from commercially-made moulds regularly advertised in wargaming journals. Casting one's armies is an eminently satisying affair, although wives tend to carp at such usurping of their traditional privilege of slaving over a hot stove! It is even more pleasing to produce one's own mould of chosen figures, knowing that they are quite unique in the world! Detailed instructions for making moulds are beyond the scope of this book, but can be obtained from such suppliers as Alec Tiranti, who regularly advertise.

Bear in mind it is both unethical and unwise to use commercial figures for this purpose; to buy a single figure and produce many from it deprives manufacturers of recompense for research and discourages them from further development.

The hobby of wargaming is most popular in Great Britain and the United States of America, with adherents in Australia and Canada, but such is its universal appeal that everything available in one country or Continent is equally valuable in another. Obviously local sources of supply are preferable to delayed overseas buying with consequent currency problems, and this can also be tackled through the pages of wargaming magazines. In America there seem to be two outstanding journals devoted to the hobby of wargaming, they are –

Courier – P.O. Box 1878, Brockton, Mass 02403, U.S.A.

Mid West Wargamers' Association Newsletter – Hal Thinglum, 22554 Pleasant Drive, Richton Park, Illinois 6047, U.S.A.

From the last available issue of each, there is a wealth of valuable information, including the names of further journals available in the United States – *The Heliograph, Savage and Soldier, P. W. Review,* and *SAGA.* Numerous model soldier manufacturers advertise or are mentioned in the pages of *Courier* and the *MWWA Newsletter,* plus retail outlets importing figures from the United Kingdom.

Not so numerous are references to sources of supply outside Great Britain and America, but can be gleaned from magazines of both countries –
AUSTRALIA – Battlefield Australia, 50 Clissold Parade, Campsie, N.S.W. Australia 2194. Retail.
    From this source it may well be possible to find out something of the Australian wargaming magazine *Breakout.*
CANADA – The Armoury, 3224 Danforth Avenue, Scarborough, Ont. MIL ICI, Canada.
GERMANY – Germania Vertrieb, Buschausener Str. 14, 4200 Oberhausen, W. Germany
J. Peddinghaus, Beethovenstrasse 20, 5870 Hemer, W. Germany.
SWEDEN – Tradition Scandinavia, P.O. Box 21170 S-100 31, Stockholm, Sweden.

Original boxes of Airfix HO/OO scale plastic figures

READ ALL ABOUT IT! THE LITERATURE OF WARGAMING

Here are most of the known books on the hobby of wargaming. Regrettably, many are out of print, but obtainable through public libraries.

Asquith, Stuart, *Guide to Solo Wargaming* (1985); *Beginner's Guide to Wargaming* (1987).

Barker, Phil, *Know the Game: Wargaming*, (1976); *Ancient Wargaming: Airfix Magazine Guide No. 9*, (1975).

Bath, Tony, *Setting Up a Wargames Campaign*, (1973).

Candler, John C., *Miniature Wargames du Temps de Napoleon*, (1964, USA).

Carter, Barry J., *Naval Wargames: World Wars 1 and 2*, (1975).

Dunn, P., *Sea Battle Games*, (1970).

Edwards ?; *Playable Napoleonic Wargames* (1987).

Featherstone, Donald, *Wargames*, (1962); *Naval War Games*, (1965); *Air War Games*, (1966); *Advanced Wargames*, (1969); *War Games Campaigns*, (1970); *Battles With Model Soldiers*, (1970); *War Games Through the Ages Vol. 1* (3000 BC–AD 1500), (1972); *War Games Through the Ages Vol. 2* (1420–1783), (1974); *War Games Through the Ages Vol. 3* (1792–1859), (1975); *War Games Through the Ages Vol. 4* (1861–1945), (1976); *Poitiers 1356 (Knights' Battles for Wargamers)*, (1972); *Battle Notes for Wargamers*, (1973); *Solo Wargaming*, (1973); *Tank Battles in Miniature; The Western Desert Campaign*, (1973); *Tank Battles in Miniature; The Mediterranean Theatre*, (1977); *Skirmish Wargames*, (1975); *Wargaming the Ancient & Medieval Periods*, (1975); *Wargaming the Pike and Shot Period*, (1977); *The Wargamers Handbook of the American Revolution*, (1977); *Battles With Model Soldiers*, (Up-dated version 1983); *Wargaming Airborne Operations*, (1977); *The Complete Wargamer's Handbook* (to be published 1988).

Featherstone, D. with Robinson, Keith, *Battles With Model Tanks*, (1979).

Grant Charles, *Napoleonic Wargaming*, (1973); *The War Game*, (1971); *The Ancient War Game*, (1974); *Battle! Practical Wargaming*, (1970); *Ancient Battles for Wargamers*, (1977).

Grant, C. S. *Wargame Tactics*, (1979); *Scenarios for Wargames*, (1981).

Gregory, Barry, *British Airborne Troops*, (McDonald & Janes 1974)

By Air To Battle – Official Account of British 1st & 6th Airborne Divisions, (Crown Copyright 1945 HMSO reprinted Patrick Stephens 1978)

Griffith, Paddy, *Napoleonic Wargaming for Fun*, (1980); *A Book of Sandhurst Wargames*, (1982). *Battle in the Civil War; Generalship & Tactics in America 1861–5*, (1987).

Gush, G., Purton, P. & Stephenson, R., *Discovering English Civil War Wargaming*, (1973).

Gush, G. with Finch, Andrew, *A Guide to Wargaming*, (1980).

Hinchcliffe, F. *Hinchcliffe Guide to Wargaming*, (1975).

Jeffrey, G. W., *The Napoleonic Wargame*, (1974).

Lyall, Gavin & Bernard, *Operation War Board; Wargaming World War Two*, (1976).

Morchauser, Joseph, *How to Play Wargames in Miniature*, (1962, USA).

Nash, David, *War Games*, (1975).

Perry, F. E., *A First Book of Wargaming*, (1977); *A Second Book of Wargaming*, (1978).

Philpott, M., & Thompson, Bob, *How to Win Wargames*, (Pamphlet).

Quarrie, Bruce, *Napoleonic Wargaming*; Airfix Magazine Guide 4, (1974); *Tank Battles in Miniature; Russian Campaign 1941–45*, (1975); *Tank Battles in Miniature; N.W. European Campaign 1944–5*, (1976); *Napoleon's Campaigns in Miniature 1796–1815*, (1977); *Tank Battles in Miniature; Arab-Israeli Wars*, (1978); *World War Two Wargaming*; Airfix Magazine Guide 15, (1976); *Beginners Guide to Wargaming* (1987).

von Reisswitz, B., *Kriegspiel* (a reprint by Bill Leeson of an 1824 book).

Sandars, John, *An Introduction to Wargaming*, (1975).

Spick, Mike, *Air Battles in Miniature*, (1978).

Taylor, Arthur, *Rules for Wargaming*, (1971).

Teague, D., *Discovering Modelling for Wargamers*, (1973).

Tugwell, Maurice, *Airborne to Battle – A History of Airborne War 1918–1971,* (Wm. Kimber London 1971)

Tunstill, John, *Discovering War Games*, (1969).

Wells, H. G., *Little Wars*, (1913).

Wesencraft, Charles, *Practical Wargaming*, (1974); *With Pike and Musket; 16th & 17th Century Battles for the Wargamer*, (1975).

Wise, Terence, *Introduction to Battle Gaming*, (1969); *Battles for Wargamers: WW2 The Western Desert*, (1972); *Battles for Wargamers: American Civil War 1862*, (1962); *Battles for Wargamers: WW2 Tunisia*, (1973); *Battles for Wargamers: The Roman Civil Wars 49–45BC*, (1974); *Battles for Wargamers: The Peninsular War 1813*, (1974); *American Civil War Wargaming*; Airfix Magazine Guide 24, (1977).

Young, Peter & Lawford, J. P., *Charge! Or How to Play Wargames*, (1969).

RULES – A NECESSARY WARGAMING EVIL

To fight a wargame one requires opposing forces, a tabletop terrain and some principles to which action or procedure conforms or is bound and intended to conform – in other words, a set of rules! Perhaps the most fascinating peculiarity about the hobby of wargaming is that few wargamers find it possible to fight amicably to rules other than those of their own devising. Possibly this is because the conditions applying in any specific wargame are almost entirely dependent upon the controlling wargamer's conception of tactics, the effect of firepower, the quality of troops and morale-effects – as reflected in the rules governing the game. The wargamer's concept of these factors arises from reading, films and hearsay so it is unlikely that two wargamers will be so balanced in their reading, assessment and critical judgement as to agree! For example, rules devoted to cavalry charges are far more likely to be realistic when their compiler actually rides horses than when the wargamer has no personal knowledge of the ability of the horse. The rules that control a wargame usually reflect the character and temperament of their devisor and a dashing wargamer will formulate rules that allow for colourful charges and encourage aggression whereas the quiet and cautious man's rules will come down strongly on the side of dour defence and punishing firepower. The wargamer, when confronted with rules he did not devise (even if commercially produced and play-tested) will adapt those rules to his own ideas and military concepts.

If rules are to provide realistic and fast-moving tabletop reconstructions of warfare then certain facts must be borne in mind. First, he must accept (albeit unwillingly) that his little metal and plastic figures have only the fighting ability and morale that he bestows upon them and, where particular soldiers are known to be of inferior morale and fighting ability, he must devise rules to reproduce

Knowledgeable rules will reflect the weight, power and potentialities of artillery of all periods of military history. Too much license will make them the 'Queen of the Battlefield'; too little will reduce them to unrealistic impotence (Courtesy Duncan Macfarlane, Wargames Illustrated)

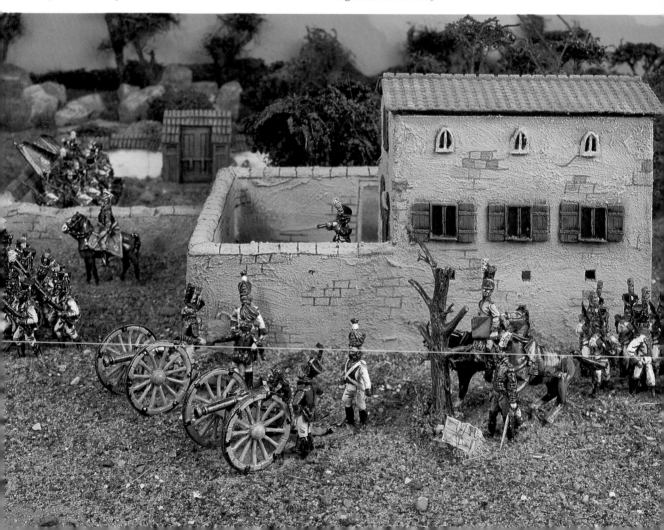

this. Many battles, fought between sturdy professional soldiers of roughly equal ability, were won or lost through disparity in the quality of leadership – this has to be covered by rules; with levies, irregulars or other inferior troops, the rules must reflect their lower fighting ability and inferior morale. Many factors of warfare are difficult to simulate on the wargames table – numerical disparity of forces, surprise, the varying power of weapons and differing states of morale, etc. Normal wargames rules do not always cover such aspects in a specific battle or campaign, so that realistic variants should be devised. On certain legendary occasions a commander took a calculated risk that came off, but this is rarely possible under normal wargames rules – would your Napoleonic rules allow Captain Ramsay's guns to successfully charge cavalry as they did at Fuentes de Onoro on 5 May 1811? Such difficult departures from the norm require rules to be slanted to accommodate them.

At some point in his career, every wargamer will want to create his own personalised set of rules, because wargamers are individualists and their direction of research and interpretation of historical facts will vary. You may, because of your readings, disagree entirely with a commercial rule publisher's ideas on the use of the lance or the breaking of a square, which spoil for you an otherwise excellent set of rules, and cause the feeling that you could do better.

There are basically two ways to formulate a personalised set of rules. The first is to write your own. This is usually the best method, and it can involve either starting from scratch, or stealing favoured bits and pieces from various published sets. Unfortunately, a set of rules compiled by a person is often looked on by himself and others as 'home-grown', and not comparable to commercial products. The second and easier course is to obtain a set of published rules you like, and modify those parts you don't like. In both cases you must be careful of the balance of the rules, and there are necessary guide-lines for achieving or maintaining balance within a set of rules.

The first determination is whether the rules are to be *strategic* or *tactical* in nature. If the rules are *strategic*, then they should involve high casualty rates, and quick and definite morale results between conflicting units, i.e. at the end of a turn, one or the other of two units in conflict should be falling back. For infantry (and possibly artillery) it should include the ability to move, or part-move, and fire at the same time. To counter this, melee results should not be weighted in favour of charg-ing infantry, nor has the defending infantry any great advantage over them. The move distance should be such that units can get into action quickly (inf. 8–12in (20–30cm) cross-country).

If the rules are to be *tactical*, infantry should only be able to move *or* fire, artillery should have move-distance deducted for limbering and unlimbering. The moves should be shorter in distance (inf. 4–8in (10–20cm) cross-country). The casualties should be considerably lower, since they are spread over more turns, and moving targets will generally be in range for longer before units can make contact. The morale should contain less severe or even partial results (eg, unit coming under fire has low morale, is forced to stop, but may return fire,) for you will be checking morale more frequently, and in a tactical game you will have opportunities to do things that could not be included in a strategic game. If your tactical rules are formulated properly, you may see the same unit charging and taking a position several times, only to get thrown out by counter-attacks each time. Melee results must be weighted heavily in favour of charging troops, for there must be a good reason of them to charge into the face of firing infantry without being able to reply. As a quick check, you should be using strategic rules if you have a wargame with over 200 figures a side that cannot remain set up overnight.

The period under consideration will affect your rules considerably. You have to take into account the weapons capabilities, tactics and training of the period as these will influence the balance of the rules. For instance, ancient archers had a fair range, but the arrow is not quite such a man-stopping device as an ounce of lead, and a lot depended on their number and deployment. Mediaeval archers generally had better bows, which gave them up to twice the range of their ancient counterpart, but they had difficulty making an impression on full plate armour – thus the rise of the shorter range but more powerful crossbow, with its armour-piercing (A.P.) quarrel. Training could have as much effect on tactics (and thus rules) as can weaponry. In the Seven Years' War, well-trained infantry could stop charging cavalry with their firepower alone, whereas Napoleonic infantry with their slightly longer-range musket were forced to form square in the face of hostile horsemen. Conversely, American Civil War infantry were able to repel cavalry, not by their superior training, but by longer-range rifled muskets; and of course, anything in a solid formation that came near a World War I infantry unit would get shot up badly. Be careful that your rules reflect the weapons and training of the time, and not the tactics. For if your rules are reasonable, the tactics of the period will be the most effective to use in your wargame under those rules.

This does not mean that one should have clauses in the rules that say, 'Infantry may only charge the enemy in column' or 'an independent battalion of skirmishers may not skirmish the whole battalion at once', or 'You may not fire into a melee'. One should never have to say that a certain tactic may or may not be used; the rules should be put together in such a way that the tactical practices that succeeded on the battlefield will also be the most useful under general conditions on the table. Under these conditions the rules are usually flexible enough to provide for alternate tactics under unusual or emergency conditions. For example, the rules can set-out the distribution of casualties between the two contenders of a melee that is fired into, plus modifications to morale; morale rules can provide that any group of skirmishers unsupported by formed troops are less inclined to stick around. So, if you need unsupported skirmishers for a short time, or want to fire into a melee, you can. If enemy cavalry are cutting-down the routed remnants of an infantry battalion and the whole mess (technically a melee) is approaching one of a battery, the gunners may fire in their own defence.

This leads to the most important point – the proper balancing of rules – movements, firepower and range, and morale, must be co-ordinated in such a way as to represent what would actually happen. For example, the weapons of the Seven Years' War were rather short range and inaccurate, but the well-disciplined troops were able to hold their fire and deliver a devastating volley at point blank range, one of which was usually sufficient to repel charging cavalry. Infantry as a rule could not come to grips; an attacking battalion would invariably be forced to stop and fire before the crunch, and whoever could least stand the musketry would have to retire. Appropriate rules would have short range but effective musketry, and morale rules that would force a unit of low morale to halt but not necessarily retire. The determination of the ability to hold fire until short range would be extremely important, especially against cavalry. Morale between conflicting infantry units may be inversely related so that as the morale of one goes up, the other's goes down, until one gets enough nerve to charge the intervening few yards just about the time the other decides to retire.

Contrast this to the less well-disciplined soldier of the Napoleonic period. Musket range is slightly increased, but the fire-power of the badly trained unit is so ineffectual that things can, and in fact have to, be decided by the bayonet. Musket range should then be just slightly longer than that of the Seven Years' War. It should not however, be so effective that it would necessarily stop charging infantry. As opposed to earlier periods, foot artillery should have a good enough move distance so that it would be advantageous to relocate it during the course of the battle.

The infantry of the American Civil War were probably no better trained or disciplined than their European predecessors of fifty years earlier, however, with their rifled musket they were able to engage the enemy at a much greater distance. Even a single line in open order could not approach closer than 40yds (36m) to steady infantry. But if frontal attacks were virtually impossible, flank attacks (if you could find an open flank) were usually assured of success. A battle line that is only one man deep and 300 to 1,000 men wide is going to be exceedingly difficult to manoeuvre. If a regiment had to change face, it had to be done one company at a time – an evolution done in battle by only the most highly disciplined troops such as the British in the Crimea. If attempted by an inexperienced unit, it usually resulted in chaos, providing an excellent target for an attacking unit.

The point is, of course, that in making up or changing rules, one should try to maintain a balance between the different elements of the rules. The movement, weapons range and fire-power, and morale should be co-ordinated in such a way that they represent the norm of the time. For instance, if an A.C.W. unit is allowed three volleys against an enemy infantry unit between the time it moves within range and the time that they come to grips (e.g. range 18in (45cm), move 6in (15cm)), and the averages obtainable for the three respective volleys are set at, say 1, 2 and 4 figures casualties, then the morale loss from taking 4 casualties in one turn, or 7 over three turns, (or better yet, both taken into account) should be more than enough to stop the attacking unit in its tracks. Conversely, if the identical situation were set in Napoleonic times, the range would be reduced to maybe 9in (23cm) (move distance 6in (15cm)) such that the defending troops could get into one, and maybe two volleys at the enemy. Then if one average (short range) volley caused 4 casualties, the morale rules should be so adjusted that, depending on the disciplines of the troops, it may or may not be enough to stop them before the crunch. Of course, the discipline of the defending troops would tend to affect its fire-power, and thus its stopping power. A Seven Years' War infantry unit should have a range about equal to their move distance, allowing one devastating volley, enough to stop the enemy.

The crux of the matter is that as you create or change certain important parts of rules systems (movement, musket range, firepower, morale) other parts must be altered to suit. For example, if you take a balanced set of rules and double the weapons range without changing any other elements of the rules, then you will double the number of shots troops can fire related to movement distance, which will have the general effect of doubling the casualties per game, doubling the chance of a morale loss making such loss twice as severe, and quartering the chance of a melee. To compensate for doubled range and maintain a balanced set of rules, you can do one of three things, in descending order of preference: you can a) halve firepower, b) double move distance, or c) cut down on the morale results.

South African wargamers in Johannesburg watch intently that the rules are observed

By the same token, doubling the move distance of a set of rules without changing aught else might provide you with the strange experience of seeing an infantry unit moving from out of range into melee with an enemy unit without a shot being fired. Or the militia unit that should have taken two steady volleys from the guard unit, takes only one, survives the single morale check (instead of two) and, to everyone's surprise, ends up crossing swords with valiant enemy legions. Obviously the weapons range should be increased to compensate.

To create the kind of rules that suit, you should first determine the average casualties in figures that you want a unit-to-unit confrontation to result in per turn. Then you must co-ordinate your missile range with your movement so that a unit will deliver a certain number of volleys per turn between the time an enemy unit comes into range and the time it comes into contact. Determine also how you want your casualties spread over the various volleys. The casualties at different ranges as an enemy infantry unit approaches should

average out to the number you determined at the beginning of this paragraph. In doing this remember that the closer the target, the more likely it is to be hit; as a famous modern General once said, the average soldier, no matter how accurate his rifle, still can't hit the broad side of a barn door past 100 paces. Rules for melees should be the same as rules for firing, for simplicity's sake; however, the results should be more casualty-producing than firepower, for the men were shooting, clubbing and stabbing each other at a range at which they could not miss.

Having got this far, test your rules out with a few very simple actions. Line two battalions of equal size up against each other and have a stationary fire-fight. Replace the casualties and do it again a couple of times. Change the range and go through it again a few more times. Then try moving a unit in column from out of range into contact with a stationary unit in line defending itself with firepower. After a few experiments of this type, you will get an impression of the average, high and low casualties. If it does not 'feel' the way you want it to, modify slightly your method of determining casualties and try it out again.

Once you have your movement and casualty determination worked out to your satisfaction, try a couple of 'fight to the last man' battles to get an idea of how basic rules work in a more complex situation. These little affairs can be quite fun and very instructive. Use equal teams that are well-balanced, including all arms – complex situations not previously considered will occur. You may also see the effect of converging fire, and the combined firepower of infantry and artillery in attack and defence. Making a comparison between these test battles and the earlier unit-to-unit confrontation, will give an idea whether or not the rules so far are working out satisfactorily. If they are not, adjust the mechanics – important before moving on to the next step.

Play-testing of basic rules before considering morale rules, will help immeasurably, giving a good idea of how hot the pace can be. You should have an idea of when a unit's morale would go bad, and what morale considerations can be based on. You will invariably come across situations where you say to yourself, 'That shouldn't happen', and will be able to make adjustments accordingly.

The mechanics of determining morale, the situations that have an influence on a unit's morale, and the results of a bad morale check, are all up to you. Build morale rules according to the ideas you like most, and they will be best for you.

Once rule-basics are squared away, extra rules can be added according to particular tastes – such subsidiary rules as will cover special weapons, special troops, special tactics, and what constitutes a charge, or an open flank, the difference in training of various troops, how long it would take to perform certain tactical manoeuvres or evolution, etc; all according to personal interpretations of researches. This is where the individuality of creating your own set of rules comes in, and the fun of it, too!

Since the range represents weapons capabilities, firepower is the troop's training, and morale their discipline, then if these things are altered to suit the times, a well written set of rules, modified accordingly, could suit any period. The only things that would need drastic change are the individual rules that apply only to that period.

The figure scale used for an army will have an effect on rules as the size of the unit and the area it covers has a decided influence both on the amount of casualties it inflicts, and receives. A 48 man battalion has twice as many men to fire (volleys), and is twice the target (ie twice as many enemy can line up opposite it) as would be a 24 man battalion. This will have a definite effect on morale results. Also one must remember that the larger a unit the more difficult it is to manoeuvre, especially in line.

A final note on wargames rules concerns the controversy arising from organised wargames – playability versus realism. Playability demands a fairly simple set of rules that might very well have to sacrifice some of the finer details to speed and simplicity. This can be handled by such physical artifices as mounting heavy cavalry 3 to a stand and giving it the same power as a larger 4 figure light cavalry stand or even larger 5 or 6 figure stands. It also generally requires opponents who know each other and get along well together, for there will invariably be certain basic parts of the rules that need to be understood (such as the exact way of moving troops, or what constitutes a flank, etc). If these things are not cleared up before the game begins, then the game may never end for the arguments.

If striving for realism, then there will be a need to continuously refer to rules and tables that differentiate between types of troops and weapons capabilities, or one will need to memorize the essence of many of these rules. The more details that are incorporated into your rules, the more various elements of your rules need to be defined, including those parts of the rules that were previously understood. However, do not fall into the false assumption that if you write a volume that defines every part of your rules, that you will eliminate arguments – for even the most carefully written definitions contain loopholes, and are subject to differing interpretations.

All games need rules and wargaming needs them more than most. Considerable reading and research are required to learn about the Armies of History and the potentialities of their weapons before accurate rules can be formulated. Many would-be wargamers attempt to cope with this by buying commercially produced sets of rules – unfortunately often they justify their claims that they realistically re-create historical conditions by being overly complicated and hard to understand.

All that is occurring on the wargames table is a battle between small model soldiers dressed as perhaps British and French, fighting a re-created or imaginative battle in their period but using tactics that would do credit to leaders in far more advanced periods of military history and are far beyond the military knowledge of the commanders. Because the little soldiers cannot march and fight, they have to be tactically directed by their war-gamer/commander, who possesses military know-ledge of a later date than the period in which he is specialising. This means that the wargamer, after researching his armies and painting them accurately so that he can proudly claim that they represent a very reasonable if scaled-down simu-lation of the army in question, is manoeuvring them with a military hindsight denied to their real-life commanders at the time in which they fought. His troops, accurate in every other respect, are performing tactical manoeuvres and evolutions far beyond the knowledge and comprehension of the men who commanded them in those far off days.

Specifically designed rules can remedy this situation by so formulating the conditions of battle that they reflect the particular attributes of the armies in question. Notwithstanding this, the war-gamer usually selects the army that most appeals to him and then uses them in a manner best calculated to bring success, but often with little relation to their historically recorded tactics. For example, the wargamer who fancies re-fighting the Peninsular War of 1808–14 will discover that the predominant tactical reason for Wellington's suc-cess over his French opponents was because he formed his well trained and confident infantry in double lines on the crest of a ridge so that they could pour a superior volume of musketry into the advancing French columns. It resembled the manner in which the French advanced in the face of English archery during the Hundred Years' War. To fight the Peninsular War in any other way but with the British in line and the French in column is to prostitute military history so as to create a com-pletely false and unreal situation.

TABLETOP BATTLEFIELDS

The numerous attractive and realistic wargames battlefields depicted in the pages of this book are ample evidence that much of the pleasure of the hobby lies in the colourful and natural appearance of the battlefields we build on our tabletops, which provide an immediate stimulus to the participants besides being a pleasure to construct. To fight a wargame without any terrain-features is like stag-ing a play against a black or single-colour backdrop – which might be avant-garde and is undoubtedly being tried in today's theatre, but it requires a good imagination to derive satisfying visual stimulation from it. Model railway hobbyists construct track-side and background scenery that is extremely realistic, being aided by a permanency denied to wargamers whose battles rarely take place on identical fields. Nevertheless, it is within the powers of the wargamer to devote time and ingenuity to constructing realistic scenic terrain. Besides pursuing authenticity in tactics, uniforms and the composition of his formations, he *must* adequately consider the terrain-factor, because when reconstructing a real-life battle, the battle-field is the most important factor and must closely resemble the historical field, both in topographical features and dimensions, otherwise what takes place upon it will bear only the most coincidental resemblance to the historical events under simu-lation. Imagine trying to fight Waterloo without the ridge, Hugoumont or La Haye Sainte; Gettysburg without the Round Tops, Cemetery Ridge or Devil's Den! It is only necessary to reproduce the actual 'fighting' areas of the battlefield so that all possible space on the wargames table is utilised, and even though it might mean 'ironing out' the known contours of the actual battlefield, all hills and slopes must be so angled as to allow model soldiers to stand up on them. If the wargame is an imaginary one perhaps forming part of a campaign then the constructed terrain must resemble the features on the campaign map.

Real life warfare has politico/economic back-ground factors which rarely intrude into our table-top battles making them largely a matter of one side or the other fighting for and gaining advan-tageous topographical positions. This means that such terrain features as hills, cross-roads, river bridges, etc, form objectives that enable wargamers to decide who has won the game and at what stage.

Terrain is more than buildings, woods and hills – it is a tactical factor that often controls the course of the battle. Few wargamers pay adequate attention to the influence of *ground* upon the tactical operations they are simulating on their tabletop battlefields, not fully appreciating that ground affects *view* and *movement* and gives *protection* to each arm whilst giving full effect to that arm. Cover from view is obtainable from quite gentle undulations and is of the greatest importance in positioning troops before serious fighting commences, or in secretly moving them from one point to another during an action, and in facilitating surprise. Cover from view does not necessarily protect from fire – hedges provide cover but not protection and should be utilised with caution because there is always a tendency to crowd men behind soft cover so that they present a dense target. On the wargames table where the wargamer towers above the field seeing everything, rules must be so formulated as to give cover its full value.

Ground affects movement by extending or limiting the front on which troops can advance and by controlling their speed over difficult surfaces. Generally speaking, troops move faster on roads than across country although roads have been known to become impassable in bad weather after numbers of troops have traversed them – it took Grouchy seven hours to move less than five miles when pursuing the Prussians from Ligny in 1815 after rain had affected the roads. Rivers and marshes seriously impede the movement of an army by imposing delay and causing changes of formation, channeling the force into a specific crossing-point. In real life such obstacles materially affect tactical operations because they make it difficult for the parts of an army to remain in communication with each other (before the days of wireless communication).

A thinking wargamer takes stock of the terrain upon which he is to fight and considers those factors which might be beneficial or detrimental to his plans. For example he will readily appreciate that terrain which tends to restrict movement is more favourable to infantry than other arms; houses, farms and villages afford advantages in defence to infantry only. If the country to the front of his force is cultivated and not too enclosed then he might consider attack to be his best policy, because his infantry will gain a succession of cover-positions allowing them to come to more equal terms with the defence. On the other hand if

A 'custom-made' model of the Waterloo building of Hougoumont

the ground in front of the enemy position is open then he will realise that defending infantry have a clear field of fire which will destroy him as he attacks.

If he is fighting in the 18th century or a period where effective musket range was short he will seek flat open ground that will permit his cavalry to approach within striking distance of infantry. But if fighting in a period where firearms are more powerful and accurate, then his cavalry must be kept at so great a distance that there will be little opportunity for them to be employed. Gone will be the thrill of glorious cavalry charges and the outcome of the battle will no longer be swayed by a powerful punch from a horsed force – now his cavalry must operate as a 'threat in being', probably only coming into their own when pursuing a retreating army. If he wishes to bring his horsed-arm within striking distance of the enemy then he must do so by means of an approach screened both from fire and view for surprise is now his only chance of success. This is done through intelligently formulated rules that eliminate the all-seeing eye of the opposing wargamer/general by means of undulating and moderately broken ground, by wooded country. Ideally, when cavalry finally come into action the ground must be as open as possible, level and free of obstacles. When a wargamer is faced with a terrain covered by hills or woods then he should leave his cavalry at home.

Artillery best come into their own on ground that is moderately undulating with long and gentle slopes, with few woods or cultivation, good roads and firm ground for wheeled movement. Ideally guns should be placed in positions that allow them to use their extensive range and give a coverless view of the enemy – but perhaps we are asking too much!

One of the most fascinating features on a wargames table is a really good hill that rolls and rises to form an admirable defensive area for one side and a point of attack for the other. Hills, in wargames as in real-life, probably provide more focal points for battles than any other feature. Some of the most famous battles in history have been fought on ridges – Hastings, Gettysburg and Waterloo to name but three. A ridge running across a table invariably seems to give a good battle. On the other hand, many battles were fought over pleasant, rolling agricultural country – particularly the mere formal 'set-piece' battles of the 18th century. Rolling country also provides interesting examples of 'dead-ground' – a very useful adjunct on the tabletop as in real life.

If they are not to dominate one's table to the exclusion of any other feature, hills must necessarily be scaled a bit smaller then they are in real life. For example, a table can be set with three or four hills each on a 16×16in (40×40cm) square; or with two larger hills on 24×24in (61×61cm) squares, or the table can carry one big hill or ridge running two-thirds across it and made up in 3 or 4 sections of 24×24in (61×61cm) that fit together to form the whole hill. Interchangeable sections can be made that will make up a variety of different hills, when rearranged.

Then there are mountains – how many of us have secret dreams of extensive campaigns amid passes, ravines, narrow tracks and trestle bridges? The essence of games in mountainous terrain lies in the fact that movement has to be made on tracks, and engagements take place on 'plateaux' or on the level ground reached after extensive mountaineering! Mountains, cliffs and bluffs also have other vital wargames uses in that, if placed on the table in clever strategical positions, they form blocks that 'channel' movement into other areas and allow for interesting tactical manoeuvres. These 'mountains' should not be too extensive as they take up valuable 'battle' space on the tabletop; rather, they should be almost flat 'symbolic' features that serve their purpose without being over-large.

The easiest way of making a permanent piece of 'hill' scenery is to nail or glue a piece of wood across the middle of a square of hardboard; cut a piece of stiff packing paper an inch larger all round than the baseboard. Glue one edge of the paper under the edge of the baseboard and stretch the paper tightly over the block of wood and glue it down on the remaining three edges. This gives a regular, rising and falling terrain that can be used with others to form rolling ground.

To make bigger hills; use an irregularly shaped baseboard; build up varying levels with pieces of wood and then cover the whole with a sheet of sacking or hessian soaked in 'Polyfilla', firmly 'bonding' to the baseboard at the edges. When dry, this will set firmly in whatever contours your wood frame allows. An even firmer affair, allowing contours to be moulded to choice, can be made by first stretching small-mesh wire-netting over the wood-blocks, then bend and push the netting into shape before laying the treated sacking over it. A similar and lighter construction can be achieved by using torn-up newspaper soaked in 'Polycel' or a similar paper-hanging adhesive instead of the sacking.

There are numerous methods of finishing off the hills so that they look realistic and attractive. They can be painted with a matt finish – 'Buckingham Green' undercoat is very effective – and then other shades of green and earth colours etc., can be blended in before the basic colour is dry. A pleasing effect can be obtained by coating the surface of the hill with an adhesive (the best for this purpose is Copydex) and then sprinkling onto it quantities of colouring powder to represent grass, earth, sand or more-or-less anything you desire. This material can be obtained at model railway shops.

Commercially made terrain pieces

Hills can be finished off by being suitably embellished with shrub (pieces of lichen moss), rocky outcrops (pieces of cork or plastic suitably painted), clumps of trees, etc, etc.

A realistic method of simulating hills, valleys and undulating ground is to stretch a green cloth or plastic sheet over carefully assembled mounds of books, slabs of polystyrene or pieces of wood. Rivers and roads can be painted onto the sheeting with poster paint and look extremely realistic.

Slabs of wood or polystyrene placed upon each other to make 'stepped' hills present a pleasing, almost 'symbolic' appearance, while providing readily definable contours and an ideal surface for soldiers to stand upon, which is the criteria.

The wargamer with a permanent room for his hobby can set up the battlefield and leave it, allowing the battle to continue for more than one session; but most wargamers have to erect and dismantle their wargames table on the same evening. Easy to assemble terrain can be made by prefabricating semi-permanent features onto chipboard squares which are fitted together on the tabletop so as to make up a variety of battlefields. If the table is 8×5ft (2.4×1.5m) then the squares should be 12×12in (30×30cm) or 2×2ft (60×60 cm) to conform; if roads and rivers emerge at the same point on each square then the squares can be permutated to make a variety of terrains. One square can carry a hill, another a farm or village; a sunken road section or a river and marsh can fill another with the squares fitted closely alongside each other so as to form perhaps a road running across the table, passing over a hill, crossing a river first by a bridge and then by a ford; at one point there is a side road winding through a wood skirting a ploughed field and leading to a farm. Very realistic scenic effects can be obtained in this way but these squares need adequate storage space or else they are damaged. Inexpensive polystyrene ceiling tiles form excellent scenic squares, being easily cut and carved into roads and rivers; hills are made by sticking shaped smaller slabs one upon the other on a large irregularly-shaped base. Polystyrene takes water colour or poster paint, although it absorbs it freely so that a thick solution is necessary.

Few stretches of land are completely bare of vegetation and the appearance of any tabletop battlefield is greatly improved by scattered clumps of natural-looking trees, bushes and scrub, which also have considerable tactical uses. These clumps, serving as a 'block' and not capable of having model soldiers actually placed within them, are easily made and most effective. Use loofah or plastic sponge cut into varying sized irregularly

A reconstructed Stalingrad for a 1942 WWII Russo-German wargame (Lionel Tarr)

rounded squares (about 3–4in) (7.5–10cm). Glue a number of these pieces onto a baseboard or on top of a hill in a rough group of varying heights. Then glue different coloured pieces of lichen-moss (obtainable from model-shops) so as to represent the tapered tops of trees emerging from the clump. Colour the whole in suitable shades; emulsion-paint or poster-paint will do but the best method is mixing powder-colour (obtainable from art shops) with a watery mixture of 'Polycel' so that a creamy liquid results. Plastered over the trees with a brush, it gives a most realistic effect.

Individual trees and bushes can be easily purchased – but if the wargamer wants to make his own then he won't do better than lichen-moss or sponge cut to shape and pierced by stout pieces of twig, the bottom of which is imbedded in 'roots' of 'plasticene'. Hedges are shaped strips of loofah sponge, painted and dotted with colour to represent hedge-plants. Marshy scrub or heathland growths can be made from small 'off-cuts' of the

Napoleon meets his Marshals outside a realistic 25mm farm built by Peter Gilder

above materials, suitably imbedded in 'plasticene' and dotted around the area like clumps of marsh grass.

Fallen trees look good; they can be made of twigs cut and trimmed to represent felled trees or else a foliage top can be cut to give the flattened, spread-out appearance of a fallen tree, the trunk (a twig) being splayed out at the bottom to look like roots, and brown 'plasticene' shaped to represent the earth that would come up with the roots.

On wargames tables, woods must be traversable by troops; this is often difficult without knocking over the trees! One simple way of getting round this is to cut out irregular shaped pieces of hardboard, paint them a dark-green and then fix three or four trees around the perimeter of the baseboard. The whole shaped board is the wood, the trees indicating this fact without getting in the way of the soldiers.

To many wargamers, particularly when they begin the hobby, roads and rivers are represented by chalked paths across the table, which doesn't look very good and they soon realise that there is considerable and effective realism in well-made roads and rivers – but they don't seem easy to construct! Strips of brown grained 'Fablon' or 'Contact' can be stuck to the actual tabletop and make realistic roads but soon lose their 'stick'. A road on the tabletop should not be too wide otherwise it takes up a disproportionate amount of

space – about 2in (5cm) for the actual road with ½in (1cm) either side for the verge is sufficient. So, cut strips of hardboard 3in (7.5cm) wide and about 24–36in (60–90cm) long; make some of them curved and include a cross-roads, T-junction, etc. Paint the roads with yellow-ochre poster paint flecked with burnt-siena for ruts, cart tracks, etc; the verges are green or can have coloured scenic powder stuck to them. These are utilitarian roads and can be laid in position and lined with hedges, walls, etc.

When using non-permanent scenery (ie when the wargames table has to be set up and dismantled for each battle), or when employing the rolling ground effect achieved by laying green plastic or cloth over blocks, it is difficult to make roads that conform to the contours, running over hill and dale. This can be done by using strips of very thin light brown plastic (used for packing), coloured realistically and laid as required, being held at either end by a piece of Blue Tack adhesive.

Road sections can be painted on baseboards large enough for the road to begin 'flat' at an edge, rise and curve, then fall again to flatness at the far edge. In this way ditches can be made by the roadside or sunken roads between high hedges etc.

Rivers can be made in much the same way as roads; and can be painted as river-sections on the reverse side of the road-strips. More elaborate rivers can be made on baseboards, then it is possible to make them fast running with clumps of rocks (cork) surrounded by white foam or with beds of rushes and marshy areas.

It is not always easy to make a river look realistic with paint. A simple and easy method is to paint your river its normal green-blue then add a number of layers of clear varnish. Another way is to paint the river, then cut strips of cellophane to size, crumple them up, straighten out and glue them to the 'river bed', the glue should be applied in areas so that the cellophane adheres patchily giving the appearance of depth and shallows.

Rivers should be 'nominal' in width and not their true scaled width, otherwise they occupy too much space that could be used for wargaming!

Many wargamers find as much, or more, pleasure in constructing the scenic features for their battle-fields than they do in the assembling and painting of their armies. Others, pushed for time, shrug the shoulders and reflect that you must have model soldiers to fight a wargame but scenery isn't vital – they don't know what they are missing!

WE'RE A SOCIABLE LOT!

It is a well-known mark of the civilised world that the community takes every opportunity of forming itself into clubs and societies, and wargamers are no different to anyone else – being probably even more gregarious than most! The majority of larger British towns and areas have flourishing wargaming clubs, some large and all-embracing (like the South London Warlords), others of a more modest nature meeting in school halls and community centres. Their number and obvious difficulties caused by frequently-changing officials make it impossible to list them in this book; however, a directory of wargames clubs can be obtained from *Military Modelling* magazine, and each issue contains details of clubs and officials.

Then there are the specific-interest groups, formed of wargamers and those interested in military history; among them are –

The Society of Ancients – magazine *Slingshot*
Society for Army Historical Research (quarterly journal)
The British Model Soldier Society (quarterly journal)
The Crimean War Research Society (magazine – *The War Correspondent*)
The British Association of Empire Players – magazine *Caisson*
The Military Historical Society (quarterly journal)
The Napoleonic Association (magazine)
The Pike and Shot Society – magazine *The Arquebusier*
The Scottish Association of Wargamers
The Seven Years' War Association – which publishes a regular newsletter
The Solo Wargamers' Association – magazine *The Lone Warrior*
The Soviet Military Research Group (Bulletin 'Red Star')
The Victorian Military Society – magazine *Soldiers of the Queen*
Wargames Developments – magazine *Nugget*
The Wessex Military Society
Undoubtedly there are many similar groups in America, among those known are –
The American Civil War Wargaming Society – magazine *Zouave*
Friends of the Foreign Legion – publishes a newsletter
The Napoleonic Society of America
The New Jersey Association of Napoleonic Wargamers – magazine *Empires, Eagles and Lions*

And no matter in what part of the world you live, a quite new dimension of tabletop wargaming will be revealed by – SKIRMISH WARGAMES (Mike Blake, The Moorings, Middle Lane, Whatstandwell, Matlock, Derbyshire, UK DE5 4RG).

Over recent years public wargaming and demonstrations of the hobby have become a recognised part of the wargaming year; shows and 'open days' have become a mandatory and leading aspect in the calendar of clubs and societies. They range from small local affairs with but a few dozen visitors to extensive and professionally organised exhibitions running over a long weekend, attracting people from distant parts as well as locals. Held throughout the year, they are initially announced and then publicised in the leading wargaming magazines of Britain and the United States of America. As a guide to what goes on, here is a list of recent conventions and shows taken from the pages of current magazines in both countries.

Great Britain
Battlegroup Open Day, Stevenage
Blitz '88, Coventry
Bridgehead '88, Beverley
Bristol and South West Model Engineers' Exhibition, Bristol
Ceasefire '88, (incl. Armoured Fighting Vehicle Association Championship), Manchester
Claymore '88, Edinburgh
Day in the Life of the 17th Century Soldier, National Army Museum, London
Essex Model Soldier Spectacular, Chelmsford
Leeds Wargames Club, F.I.A.S., Leeds
Milton Keynes Wargames Society, Annual Convention, Milton Keynes
Model Show, Plumpton Racecourse, Sussex
Modelville '88, Portsmouth
National Wargame Championship, Nottingham
Northern Militaire, Manchester
Northern Model Show, Newcastle
Reveille '88, Bristol
Sabre '88, Harrowgate
Salute '88, Kensington, London
Sensation Convention 1988, Glasgow
South Beds Immortals, Dunstable
Spring Militaire, Swinton, Manchester
Sword and Lance, Darlington
Tunbridge Wells Wargames Society Open Day, Southborough, Kent
Valhalla '88, Farnborough
Vietnam Teach-In, Wolverhampton
West Midland Military Modelling Show, Walsall
Overseas
British Forces Germany Wargaming Convention, Rheindahlen, B.A.O.R.

'Arms and Armour Press' Waterloo Convention at a London hotel on the 160th Anniversary of the Battle in 1975. Terrain and figures by Mike Willmore (featured elsewhere in this book)

Migs IX Firestone War Veterans' Association, Hamilton, Ont. Canada

USA

Annapolis Toy Soldier Show, Annapolis

Atlanticon's ORIGINS, Historicon '88, Baltimore

Historical Miniatures Games Day, Medford, Ma

Milwaukee Wargaming Weekend

Northern Virginia Adventure Gamers Third Annual Miniature Gaming Convention, Leesburg, Virginia

Outdated even before being written here is a Calender of Events published in a recent issue of the American wargaming magazine *Courier*, indicating the well organised and widespread nature of this attractive aspect of the hobby.

Oct 2–4, 1987 – Suncoast Skirmishes '87, Howard Johnson Plaza Hotel Tampa, FL.

Oct 2–4, 1987 – Council of Five Nations Con, Centre City Skating Rink, Schenectady, NY

Oct 3–4, 1987 – IXth Historical Weekend, S.Milwaukee, WI

Oct 3–4, 1987 – Autumn Campaigns '87, Lexington, KY

Oct 3–4, 1987 – Toledo Gaming Convention, Toledo University, Scottpark Campus, Toledo, Ohio

Oct 16–18, 1987 – Rudicon 3, Rochester Institute of Technology, Rochester, NY

Oct 17–18, 1987 – Council Fires III, Paris Golf and Country Club, Paris, Ont., Canada

Oct 24–25, 1987 – Novag III, Westpark Hotel, Leesburg, VA

Nov 7–8, 1987 – Rock-Con XIV, Wagon Wheel Resort, Rockton, IL

Nov 14, 1987 – A Gathering of the Clans, VFW Hall, Mystic Aven, Medford, MA

Feb 12–14, 1988 – Genghis Con IX, Airport Hilton, Denver, CO

March 4–6, 1988 – Bashcon '88, Toledo, OH

March 4–6, 1988 – Jaxcon South 12, Jacksonville Hotel on the Riverwal, Jacksonville, FL

Undoubtedly there are others not listed, and a similar diary for Great Britain would reveal an astonishing number of such public meetings held throughout the year in all four corners of the country. Check in the pages of one of the magazines, then go along and meet others with similar interests to your own – you'll be hooked!

LET'S FORM A CLUB!

Mankind is endowed with a peculiar trait which makes him desire to congregate together with his fellows. The English-speaking races translate this natural gregariousness, this herd instinct into reality by getting together to form clubs of people with similar interests, be they stamp collecting or even mass drinking! Perhaps defensively it can be claimed that the wargamer establishes and joins clubs together with his fellow hobbyist for a more tangible and constructive purpose. In the first place, wargaming is a hobby that obviously is more suited to the joint efforts of two or more people rather than a single individual ploughing a lonely furrow. This is because it is competitive and because much of what occurs on the wargames table and the campaigns map is essentially a battle of wits in which secrecy plays no small part. For these reasons alone, the wargamer is encouraged to seek out at least one other person of similar interests.

On top of that, there is no doubt whatsoever that in many cases a combination of new ideas and resources of a number of people add greatly to the enjoyment and true depth of the hobby. With the ever increasing range of readily available wargames figures the massing of armies is no longer an expensive business but it still takes an immeasurably long time to paint up sufficient figures to form two large enough forces to give a realistic battle or campaign. By joining with his fellows, each collecting and painting to a set plan, the wargamer is soon able to take part in large-scale battles involving far more model soldiers than he could paint in a year.

There are other benefits such as those gained by the man or boy who does not have the necessary space to lay out wargames tables of any great size or indeed to lay them out at all. By joining a Club such a person is able to take advantage of tabletop terrains set out in a variety of periods and to a standard that may well surpass anything he could himself achieve. Then there is the matter of variety – most wargamers have a favourite period in which they collect their armies. Often it is the Napoleonic period when they attempt to build up armies of British and French, taking all their available cash and time in so doing. That same man, whilst his enthusiasm for the Napoleonic era will never wane, may also have secret or not so secret yearnings to fight battles with chariots, war elephants and Roman legions or else he may desire to try his hand at modern warfare with tanks and the other multi-weaponed forces of World War II. By joining a

Wargames Club he is able to indulge himself in these fancies and in return his figures can be used on Club nights by a modernist or an Ancient-fan who has always wanted to be a Wellington or a Napoleon.

The well-organised Wargames Club meets at least one evening a month or ideally on a Saturday or Sunday afternoon; and probably more often. The first arrivals set up 6ft by 2ft (2×0.6m) trestle tables and begin to lay out terrain they have brought with them or else taken from the Club collection donated by members and left on the premises. When the later arrivals turn up they have the choice of fighting in perhaps any one of five or six periods so that in the course of a year a young wargamer can sample the delights of battling in many areas before finally selecting that which appeals most to him.

Some Clubs are more fortunate than others and instead of hiring the village hall they are able to secure some permanent premises in which they can safely store their terrain and models, leave large-scale battles in situ from week to week and have far more frequent meetings than are possible when hiring fees have to be paid. It might be a cellar beneath a member's house which has to be cleaned out and turned into a Club room or an old Nissen hut set on a piece of waste land – although in this age of the permissive society it is feared that intruders might well viciously destroy such items as model soldiers and terrain which are of no value to them but of inestimable value to their owners. A permanent Club room also allows the Club to have a small library or an information service, with members providing spare books from their own collection. Joint subscriptions can be taken to magazines of interest to wargamers and military collectors which can be shared around the Club on a rota basis.

Perhaps the greatest boon of a Wargames Club lies in the many projects made possible by its larger numbers of members and their joint facilities. Usually, as already mentioned, a wargames evening finds a number of tables in varying periods set up and awaiting combatants. Inevitably there comes the week when holidays intervene or human nature takes over and everyone leaves it to everyone else to bring something along so that games have to be played with very small numbers, and hasty journeys home are made to bring back enough troops to set up a table. This can be avoided by having a Club Project that runs throughout the year in addition to the other activities of the Club. Such a project should be carefully chosen after consultation in which all members of the Club have a voice; it should be made mutually interesting and possess a built-in factor that allows its competitive interests to run from month to month over a long period without any diminishing of interest. Perhaps the simplest way of tackling such a situation is to have a large map made up of perhaps a dozen wargames tables. Two commanders-in-chief are selected, who deploy their forces on the map and then co-ordinate the results of each battle and carry on accordingly from month to month. The method of fighting is for one of the wargames tables to be set up at each meeting of the Club and for a specified general (each member of the Club takes a turn at this) to command a force under the direct orders of his C-in-C. At the end of the battle the result, together with losses, etc, is handed to each commander-in-chief and to the umpire who marks them up on his map. At the next meeting another one of the wargames tables is constructed and a second general fights a battle in that area. He gives his results and losses, etc, to his c-in-c and so the procedure goes on month by month. The simple truth about such a project is that the commanders-in-chief have perhaps 10 regiments at their disposal in all, but as those 10 regiments are used in one sector in the first week and the same 10 in another sector in the second meeting and so on, it is possible for 10 regiments to be multiplied by say 12 monthly battles so that each commander-in-chief commands (on paper) a force of 120 regiments which are deployed throughout his map. This is a simple method of 'audience-participation' – a factor which should be the aim of every Wargames Club because it is the kiss of death to have young members or shy members coming along to meetings and standing by without taking part or being welcomed into any game.

One well-known American wargamer regularly presides over very large-scale games, usually reconstructions of actual battles of the Napoleonic Wars. These games involve as many as 20 or 30 wargamers, each of whom has a command and who receives his battle orders and then fights accordingly. These commands are graded according to the experience of the wargamer who starts at the bottom with a small command and progresses slowly up the ladder of promotion until he commands a Division or Corps.

Perhaps it is not entirely to the taste of the conservative English wargamer to actually bear a military rank in the Club and to seriously accept promotion or demotion according to his prowess on the wargames table. Nevertheless every man has a competitive bone in his body so that the stimulus and interest of rising or falling in some published League Table has an appeal. If Club Projects involve certain designated regiments or formations provided by individual wargamers, so that Charlie Jones is well known as the brave commander of the 37th Regiment of Foot, the 88th Regiment of Foot and the Scots Greys, then any battle honours gained by those regiments will add lustre to the name of Charlie Jones. The manner in which this can be demonstrated is for the Club artist to paint a series of small outlines of flags on a large sheet of cartridge paper. Each regiment will have its own flag on which will be marked those battles at which it distinguishes itself.

To get a Wargames Club off the ground is not an easy matter – to arouse a 'Club spirit' or sense of belonging to a group of individuals of varying ages and occupations requires a focal aim to serve as a banner on which all can be rallied. A novel method with a merit of being inexpensive can be the formation of a Club wargames army. First every member of the Club is asked to donate the price of a box of inexpensive plastic figures, say ESCI or Airfix – and as an example let us assume that sixteen sums have been collected. Hold a meeting at which the project is explained, when everyone is told that the Club are going to buy sixteen boxes of figures, and that each member will be responsible for painting up one box of figures which will then become the Club property and will always be available to be used on wargames evenings. If there are no strong objections perhaps the easiest period to take is the American Civil War for which four boxes of Federal Infantry, two boxes of Cavalry and two boxes of Artillery are purchased. The same numbers are purchased for the Confederate forces. Everyone knows that these boxes of figures contain about forty figures or so in a large variety of positions, by taking all the men firing and placing them together and all the men who are running, and all the men kneeling, etc, etc, one can obtain regiments of say twenty men all in firing positions or forty men all in kneeling positions together with officers and standard bearers. Divide the forces up in this manner and allocate one regiment to each man who is then responsible for painting it and bringing it along to the Club. On the duly designated evening, if everyone has played up to form, two small armies of Federals and Confederates will be ready for action. In addition to the fact that everyone is made to feel that they are part of the project, there is also the interest in comparing styles of painting and perhaps, in battles that follow, proudly handling or following the fortunes of a regiment painted by oneself.

It is particularly applicable to employ this specific war as a basic Club Project because it is extensively considered in Chapter 2, thus obviating the necessity for inexperienced players to seek around for essential details. Also, the chapter concludes with a simple set of rules easily comprehended by the merest novice and ideal for such a Club Project. An experienced member is required to act as Campaign Co-ordinator who will organise each battle, determine the terrain and have it set out on the day – using whatever battlefield scenery the Club has available. Also, he will set down conditions for the specific battle ie initial laying-out positions for each army, and the objectives of each commander – thus one battle can be a straightforward 'laying-out on each baseline and advance to contact' or one force has to hold a hill or village; or perhaps a skilful retreat with minimal losses, etc, etc.

A preliminary meeting will discuss the project, when those wishing to participate declare themselves for the Union (Federals) or the Confederacy, thus pledging himself (or herself) to assemble a small personal force to be brought along to each battle. Such forces – on economic and practical grounds – are restricted to a maximum of *two* regiments of infantry and one gun and crew, or *two* cavalry units plus a gun, or *one* foot regiment, *one* cavalry unit and *one* gun. A convenient facet of the rules given in Chapter 2 is that the actual size (ie numbers of men) of a regiment is unimportant, because the unit fights and reacts as a whole and not as individuals being removed when they become casualties.

And it all really works! These suggestions are not given 'off the top of the head' but have been extensively 'play-tested' over many months at regular meetings of the Wessex Wargaming Society – admittedly way back in the nostalgic past, but one of the beauties of wargaming lies in its agelessness!

The way in which the campaign was conducted on the evenings when we met to fight held several novel points. In the first place the Supreme Commander of each army, ie, the General, was selected on the night of the battle when a draw was made between all those taking part on each side. The man so drawn as commander could not command any actual force in the forthcoming battle and had to hand his regiments over to another. He commanded through junior commanders whom he selected from those on his side and he dealt only with them and not with unit commanders. At the conclusion of each battle, the general signed a proforma showing how the battle developed. This was

John Cox shows a young club member how to solder

sent to the campaign co-ordinator who kept records and scores.

In order that battles could be fought to a finish in an evening, no battle was to last longer than eight game-moves. At that point, night was deemed to have fallen and an assessment was made as to whom the situation favoured. An interesting and realistic situation was built into the campaign in that on the night of a battle the Federal forces were represented only by those wargamers who turned up with troops of that army – similarly the armies of the Confederacy were represented by wargamers turning up with their Johnny Rebs so that, just as in a rather muddled Civil War like the one we were attempting to simulate, battles took place between whoever happened to get to the field at the right time. Needless to say there was a considerable amount of lobbying and pressure put upon missing members to turn up and swell the numbers. To allow for the fact that one side might be larger than the other, the battle instructions of the Co-ordinator were flexible in that he detailed a situation in which a smaller force defended against a larger force. On the night, if more had turned up with Federal troops than with Confederates, then the Federals were the larger force and did the attacking, whilst Robert E. Lee with the smaller army did the defending.

The routine that came to be accepted on wargames nights was for everyone to assemble at the appointed time when the respective supporters of both sides declared themselves and took their post on either side of the wargames table to cry out challenges and abusive remarks across the table. Next the Club secretary opened the sealed instructions of the campaign co-ordinator and they were read aloud. Until this point no one had any knowledge of the instructions. Then the terrain was set up in accordance with details contained in the instructions and the respective generals were drawn for. They selected their junior commanders and proceeded to work out their tactics. Then the forces were laid out and the battle commenced. Eight moves later the result was decided, both commanders completed and signed the pro-forma and then everyone retired round the corner to a handy pub to re-fight the battle.

It turned out to be everything that was hoped for with at least twenty members purchasing and painting up armies for one side or the other. On the night of the battle there were never less than fourteen or sixteen players, each of whom handled his own small group on the battlefield and felt justifiably proud or ashamed of their progress.

Down at the Wargames Club – a battle in progress

Points were scored as a result of each battle and eventually a League Table was made out showing the respective positions of the six armies. Scoring was as follows:

For achieving objective – 10 points
A drawn battle – 5 points
For each standard captured – 3 points
For each General killed or captured – 5 points
For each junior commander killed or captured – 3 points
For each gun captured – 2 points
For causing enemy 25% casualties – 2 points
For causing enemy 50% casualties – 5 points
For causing enemy 75% casualties – 7 points

This very successful Club Project did much to get the Club 'off the ground' through sheer enthusiasm generated by making members take sides and support them. At times it all got a bit serious and even heated, resembling an election campaign because, in a quite unpremeditated manner, those members with Labour Party sympathies gravitated towards the Union, whilst the Confederate supporters favoured the Conservative Party!

INDEX